W9-BWX-412

Gordon College
Wenham, Mass. 01984

DISCARDED
JENKS LRC
GORDON COLLEGE

THE DOUBLE-EDGED HELIX
Science in the Real World

CONVERGENCE

Founded, Planned, and Edited by
RUTH NANDA ANSHEN

THE DOUBLE-EDGED HELIX
Science in the Real World

LIEBE F. CAVALIERI

1981
COLUMBIA UNIVERSITY PRESS
NEW YORK

Q
175
.C435

Library of Congress Cataloging in Publication Data

Cavalieri, Liebe F., 1919–
The double-edged helix.

(Convergence; v. 2)
Bibliography: p.
Includes index.
1. Science—Philosophy. 2. Man. 3. Philosophy
of nature. 4. Cosmology. I. Title. II. Series.
Q175.C435   501   81-2213
ISBN 0-231-05306-1   AACR2

Columbia University Press
New York   Guildford, Surrey

Copyright © 1981 by Liebe F. Cavalieri
Introductory Essays, "Convergence" and "The Möbius Strip,"
Copyright © 1981 by Ruth Nanda Anshen

All rights reserved
Printed in the United States of America

Printed on permanent and durable acid-free paper

Contents

Convergence   Ruth Nanda Anshen   **1**

The Möbius Strip   R.N.A.   **13**

Preface   **15**

1. Dilemmas   **20**

2. A Scientist Looks at Science: An Overview   **24**

3. Gene-Splicing   **41**

4. The Hazards of Success   **58**

5. Science as Technology, and Vice Versa   **71**

6. Rousseau Revisited   **84**

7. From Truth to Power   **102**

8. Freedom of Inquiry   **128**

9. Conscience in Science   **142**

References and Notes   **158**

Index   **189**

About the Author   **195**

About the Founder of this Series   **196**

Already Published in

CONVERGENCE

EMERGING COSMOLOGY   *Bernard Lovell*

THE DOUBLE-EDGED HELIX
Science in the Real World   *Liebe F. Cavalieri*

Board of Editors
of
CONVERGENCE

Liebe F. Cavalieri
Erwin Chargaff
Sir Fred Hoyle
Sir Bernard Lovell
Adolph Lowe
Joseph Needham
I. I. Rabi
Karl Rahner
William Richardson
Jonas Salk
Lewis Thomas
C. N. Yang

Convergence
by
Ruth Nanda Anshen

"There is no use trying," said Alice; "one *can't* believe impossible things."

"I dare say you haven't had much practice," said the Queen. "When I was your age, I always did it for half an hour a day. Why, sometimes I've believed as many as six impossible things before breakfast."

This commitment is an inherent part of human nature and an aspect of our creativity. Each advance of science brings increased comprehension and appreciation of the nature, meaning and wonder of the creative forces that move the cosmos and created man. Such openness and confidence lead to faith in the reality of possibility and eventually to the following truth: "The mystery of the universe is its comprehensibility."

When Einstein uttered that challenging statement, he could have been speaking about our relationship with the universe. The old division of the Earth and the Cosmos into objective processes in space and time and mind in which they are mirrored is no longer a suitable starting point for understanding the universe, science, or ourselves. Science now begins to focus on the convergence of man and nature, on the framework which makes us, as living beings, dependent parts of nature and simultaneously makes nature the object of our thoughts and actions. Scientists can no longer confront the universe as objective observers. Science recognizes the participation of

1

man with the universe. Speaking quantitatively, the universe is largely indifferent to what happens in man. Speaking qualitatively, nothing happens in man that does not have a bearing on the elements which constitute the universe. This gives cosmic significance to the person.

Our hope is to overcome the cultural *hubris* in which we have been living. The scientific method, the technique of analyzing, explaining, and classifying, has demonstrated its inherent limitations. They arise because, by its intervention, science presumes to alter and fashion the object of its investigation. In reality, method and object can no longer be separated. The outworn Cartesian, scientific world view has ceased to be scientific in the most profound sense of the word, for a common bond links us all—man, animal, plant, and galaxy—in the unitary principle of all reality. For the self without the universe is empty.

This universe of which we human beings are particles may be defined as a living, dynamic process of unfolding. It is a breathing universe, its respiration being only one of the many rhythms of its life. It is evolution itself. Although what we observe may seem to be a community of separate, independent units, in actuality these units are made up of subunits, each with a life of its own, and the subunits constitute smaller living entities. At no level in the hierarchy of nature is independence a reality. For that which lives and constitutes matter, whether organic or inorganic, is dependent on discrete entities that, gathered together, form aggregates of new units which interact in support of one another and become an unfolding event, in constant motion, with ever-increasing complexity and intricacy of their organization.

Are there goals in evolution? Or are there only discernible patterns? Certainly there is a law of evolution by which we can explain the emergence of forms capable of activities which are indeed novel. Examples may be said to be the origin of life, the emergence of individual consciousness, and the appearance of language.

The hope of the concerned authors in CONVERGENCE is

that they will show that evolution and development are interchangeable and that the entire system of the interweaving of man, nature, and the universe constitutes a living totality. Man is searching for his legitimate place in this unity, this cosmic scheme of things. The meaning of this cosmic scheme —if indeed we can impose meaning on the mystery and majesty of nature—and the extent to which we can assume responsibility in it as uniquely intelligent beings, are supreme questions for which this Series seeks an answer.

Inevitably, toward the end of a historical period, when thought and custom have petrified into rigidity and when the elaborate machinery of civilization opposes and represses our more noble qualities, life stirs again beneath the hard surface. Nevertheless, this attempt to define the purpose of CONVERGENCE is set forth with profound trepidation. We are living in a period of extreme darkness. There is moral atrophy, destructive radiation within us, as we watch the collapse of values hitherto cherished—but now betrayed. We seem to be face to face with an apocalyptic destiny. The anomie, the chaos, surrounding us produces an almost lethal disintegration of the person, as well as ecological and demographic disaster. Our situation is desperate. And there is no glossing over the deep and unresolved tragedy that fills our lives. Science now begins to question its premises and tells us not only what *is,* but what *ought* to be; *pre*scribing in addition to *de*scribing the realities of life, reconciling order and hierarchy.

My introduction to CONVERGENCE is not to be construed as a prefatory essay to each individual volume. These few pages attempt to set forth the general aim and purpose of this Series. It is my hope that this statement will provide the reader with a new orientation in his thinking, one more specifically defined by these scholars who have been invited to participate in this intellectual, spiritual, and moral endeavor so desperately needed in our time. These scholars recognize the relevance of the nondiscursive experience of life which the discursive, analytical method alone is unable to convey.

The authors invited to CONVERGENCE Series acknowledge a structural kinship between subject and object, between living and nonliving matter, the immanence of the past energizing the present and thus bestowing a promise for the future. This kinship has long been sensed and experienced by mystics. Saint Francis of Assisi described with extraordinary beauty the truth that the more we know about nature, its unity with all life, the more we realize that we are one family, summoned to acknowledge the intimacy of our familial ties with the universe. At one time we were so anthropomorphic as to exclude as inferior such other aspects of our relatives as animals, plants, galaxies, or other species—even inorganic matter. This only exposed our provincialism. Then we believed there were borders beyond which we could not, must not, trespass. These frontiers have never existed. Now we are beginning to recognize, even take pride in, our neighbors in the Cosmos.

Human thought has been formed through centuries of man's consciousness, by perceptions and meanings that relate us to nature. The smallest living entity, be it a molecule or a particle, is at the same time present in the structure of the Earth and all its inhabitants, whether human or manifesting themselves in the multiplicity of other forms of life.

Today we are beginning to open ourselves to this evolved experience of consciousness. We keenly realize that man has intervened in the evolutionary process. The future is contingent, not completely prescribed, except for the immediate necessity to evaluate in order to live a life of integrity. The specific gravity of the burden of change has moved from genetic to cultural evolution. Genetic evolution itself has taken millions of years; cultural evolution is a child of no more than twenty or thirty thousand years. What will be the future of our evolutionary course? Will it be cyclical in the classical sense? Will it be linear in the modern sense? Certainly, life is more than mere endless repetition. We must restore the importance of each moment, each deed. This is impossible if the future is nothing but a mechanical extrapolation of the past. Dignity becomes possible only with choice. The choice is ours.

In this light, evolution shows man arisen by a creative power inherent in the universe. The immense ancestral effort that has borne man invests him with a cosmic responsibility. Michelangelo's image of Adam created at God's command becomes a more intelligent symbol of man's position in the world than does a description of man as a chance aggregate of atoms or cells. Each successive stage of emergence is more comprehensive, more meaningful, more fulfilling, and more converging, than the last. Yet a higher faculty must always operate through the levels that are below it. The higher faculty must enlist the laws controlling the lower levels in the service of higher principles, and the lower level which enables the higher one to operate through it will always limit the scope of these operations, even menacing them with possible failure. All our higher endeavors must work through our lower forms and are necessarily exposed thereby to corruption. We may thus recognize the cosmic roots of tragedy and our fallible human condition. And language itself as the power of universals, is the basic expression of man's ability to transcend his environment and to transmute tragedy into a moral and spiritual triumph.

This relation of the higher to the lower applies again when an upper level, such as consciousness or freedom, endeavors to reach beyond itself. If no higher level can be accounted for by the operation of a lower level, then no effort of ours can be truly creative in the sense of establishing a higher principle not intrinsic to our initial condition. And establishing such a principle is what all great art, great thought, and great action must aim at. This is indeed how these efforts have built up the heritage in which our lives continue to grow.

Has man's intelligence broken through the limits of his own powers? Yes and no. Inventive efforts can never fully account for their success, but the story of man's evolution testifies to a creative power that goes beyond that which we can account for in ourselves. This power can make us surpass ourselves. We exercise some of it in the simple act of acquiring knowledge and holding it to be true. For, in doing so, we strive for intellectual control over things outside ourselves, in spite of our manifest incapacity to justify this hope. The greatest efforts of the human mind amount to no more than this. All such acts impose an obligation to strive for the ostensibly impossible, representing man's search for the fulfillment of those ideals which, for the moment, seem to be beyond his reach.

The origins of one person can be envisaged by tracing that person's family tree all the way back to the primeval specks of protoplasm in which his first origins lie. The history of the family tree converges with everything that has contributed to the making of a human being. This segment of evolution is on a par with the history of a fertilized egg developing into a mature person, or the history of a plant growing from a seed; it includes everything that caused that person, or that plant, or that animal, or even that star in a galaxy, to come into existence. Natural selection plays no part in the evolution of a single human being. We do not include in the mechanism of growth the possible adversities which did not befall it and hence did not prevent it. The same principle of development holds for the evolution of a single human being; nothing is gained in understanding this evolution by considering the adverse chances which might have prevented it.

In our search for a reasonable cosmic view, we turn in the first place to common understanding. Science largely relies for its subject matter on a common knowledge of things. Concepts of life and death, plant and animal, health and sickness, youth and age, mind and body, machine and technical processes, and other innumerable and equally important things are commonly known. All these concepts apply to complex entities, whose reality is called into question by a theory of knowledge which claims that the entire universe should ultimately be represented in all its aspects by the physical laws governing the inanimate substrate of nature.

Our new theory of knowledge, as the authors in this Series try to demonstrate, rejects this claim and restores our respect for the immense range of common knowledge acquired by our experience of convergence. Starting from here, we sketch out our cosmic perspective by exploring the wider implications of the fact that all knowledge is acquired and possessed by relationship, coalescing, merging.

We identify a person's physiognomy by depending on our awareness of features that we are unable to specify, and this amounts to a convergence in the features of a person for the purpose of comprehending their joint meaning. We are also able to read in the features and behavior of a person the presence of moods, the gleam of intelligence, the response to animals or a sunset or a fugue by Bach; the signs of sanity, human responsibility, and experience. At a lower level, we comprehend by a similar mechanism the body of a person and understand the functions of the physiological mechanism. We know that even physical theories constitute in this way the processes of inanimate nature. Such are the various levels of knowledge acquired and possessed by the experience of convergence.

The authors in this Series grasp the truth that these levels form a hierarchy of comprehensive entities. Inorganic matter is comprehended by physical laws; the mechanism of physiology is built on these laws and enlists them in its service. Then, the intelligent behavior of a person relies on the healthy functions of the body and, finally, moral responsibility relies on the faculties of intelligence directing moral acts.

We realize how the operations of machines, and of mechanisms in general, rely on the laws of physics but cannot be explained, or accounted for, by these laws. In a hierarchic sequence of comprehensive levels, each higher level is related to the levels below it in the same way as the operations of a machine are related to the particulars, obeying the laws of physics. We cannot explain the operations of an upper level in terms of the particulars on which its operations rely. Each higher level of integration represents, in this sense, a higher level of existence, not completely accountable by the levels below it yet including these lower levels implicitly.

In a hierarchic sequence of comprehensive levels each higher level is known to us by relying on our awareness of the particulars on the level below it. We are conscious of each level by internalizing its particulars and mentally performing the integration that constitutes it. This is how all experience, as well as all knowledge, is based on convergence, and this is how the consecutive stages of convergence form a continuous transition from the understanding of the inorganic, the inanimate, to the comprehension of man's moral responsibility and participation in the totality, the organismic whole, of all reality. The sciences of the subject-object relationship thus pass imperceptibly into the metascience of the convergence of the subject and object interrelationship, mutually altering each other. From the minimum of convergence, exercised in a physical observation, we move without a break to the maximum of convergence, which is a total commitment.

"The last of life, for which the first was made, is yet to come." Thus, CONVERGENCE has summoned the world's most concerned thinkers to rediscover the experience of *feeling*, as well as of thought. The convergence of all forms of reality presides over the possible fulfillment of self-awareness—not the isolated, alienated self, but rather the participation in the life process with other lives and other forms of life. Convergence is a cosmic force and may possess liberating powers allowing man to become what he is, capable of freedom, justice, love. Thus man experiences the meaning of grace.

A further aim of this Series is not, nor could it be, to disparage science. The authors themselves are adequate witness to this fact. Actually, in viewing the role of science, one arrives at a much more modest judgment of its function in our whole body of knowledge. Original knowledge was probably not acquired by us in the active sense; most of it must have been given to us in the same mysterious way we received our consciousness. As to content and usefulness, scientific knowledge is an infinitesimal fraction of natural knowledge. Nevertheless, it is knowledge whose structure is endowed with beauty because its abstractions satisfy our urge for specific knowledge much more fully than does natural knowledge, and we are justly proud of scientific knowledge because we can call it our own creation. It teaches us clear thinking, and the extent to which clear thinking helps us to order our sensations is a marvel which fills the mind with ever new and increasing admiration and awe. Science now begins to include the realm of human values, lest even the memory of what it means to be human be forgotten.

No individual destiny can be separated from the destiny of the universe. Alfred North Whitehead has stated that every event, every step or process in the universe, involves both effects from past situations and the anticipation of future potentialities. Basic for this doctrine is the assumption that the course of the universe results from a multiple and never-ending complex of steps developing out of one another. Thus, in spite of all evidence to the contrary, we conclude that there is a continuing and permanent energy of that which is not only man but all of life. For not an atom stirs in matter, organic and inorganic, that does not have its cunning duplicate in mind. And faith in the convergence of life with all its multiple manifestations creates its own verification.

We are concerned in this Series with the unitary structure of all nature. At the beginning, as we see in Hesiod's *Theogony* and in the Book of Genesis, there was a primal unity, a state of fusion in which, later, all elements become separated but then merge again. However, out of this unity there emerge, through separation, parts of opposite elements. These opposites intersect or reunite, in meteoric phenomena or in individual living things. Yet, in spite of the immense diversity of creation, a profound underlying convergence exists in all nature. And the principle of the conservation of energy simply signifies that there is a *something* that remains constant. Whatever fresh notions of the world may be given us by future experiments, we are certain beforehand that something remains unchanged which we may call *energy*. We now do not say that the law of nature springs from the invariability of God, but with that curious mixture of arrogance and humility which scientists have learned to put in place of theological terminology, we say instead that the law of conservation is the physical expression of the elements by which nature makes itself understood by us.

The universe is our home. There is no other universe than the universe of all life including the mind of man, the merging of life with life. Our consciousness is evolving, the primordial principle of the unfolding of that which is implied or contained in all matter and spirit. We ask: Will the central mystery of the cosmos, as well as man's awareness of and participation in it, be unveiled, although forever receding, asymptotically? Shall we perhaps be able to see all things, great and small, glittering with new light and reborn meaning, ancient but now again relevant in an iconic image which is related to our own time and experience?

The cosmic significance of this panorama is revealed when we consider it as the stages of an evolution that has achieved the rise of man and his consciousness. This is the new plateau on which we now stand. It may seem obvious that the succession of changes, sustained through a thousand million years, which have transformed microscopic specks of protoplasm into the human race, has brought forth, in so doing, a higher and altogether novel kind of being, capable of compassion, wonder, beauty and truth, although each form is as precious, as sacred, as the other. The interdependence of everything with everything else in the totality of being includes a participation of nature in history and demands a participation of the universe.

The future brings us nothing, gives us nothing; it is we who in order to build it have to give it everything, our very life. But to be able to give, one has to possess; and we possess no other life, no living sap, than the treasures stored up from the past and digested, assimilated, and created afresh by us. Like all human activities, the law of growth, of evolution, of convergence draws its vigor from a tradition which does not die.

CONVERGENCE is committed to the search for the deeper meanings of science, philosophy, law, morality, history, technology, in fact all the disciplines in a transdisciplinary frame of reference. This Series aims to expose the error in that form of science which creates an unreconcilable dichotomy between the observer and the participant, thereby destroying the uniqueness of each discipline by neutralizing it. For in the end we would know everything but *understand nothing,* not being motivated by concern for any question. This Series further aims to examine relentlessly the ultimate premises on which work in the respective fields of knowledge rest and to break through from these into the universal principles which are the very basis of all specialist information. More concretely, there are issues which wait to be examined in relation to, for example, the philosophical and moral meanings of the models of modern physics, the question of the purely physico-chemical processes versus the postulate of the irreducibility of life in biology. For there is a basic correlation of elements in nature, of which man is a part, which cannot be separated, which compose each other, which converge, and alter each other mutually.

Certain mysteries are now known to us: the mystery, in part, of the universe and the mystery of the mind have been in a sense revealed out of the heart of darkness. Mind and matter, mind and brain, have converged; space, time, and motion are reconciled; man, consciousness, and the universe are reunited since the atom in a star is the same as the atom in man. We are homeward bound because we have accepted our convergence with the Cosmos. We have reconciled observer and participant. For at last we know that time and space are modes by which we think, but not conditions in which we live and have our being. Religion and science meld; reason and feeling merge in mutual respect for each other, nourishing each other, deepening, quickening, and enriching our experiences of the life process. We have heeded the haunting voice in the whirlwind.

The Möbius Strip

The symbol found on jacket and binding of each volume in Convergence is the visual image of *convergence*—the subject of this Series. It is a mathematical mystery deriving its name from Augustus Möbius, a German mathematician who lived from 1790 to 1868. The topological problem still remains unsolved mathematically.

The Möbius Strip has only one continuous surface, in contrast to a cylindrical strip, which has two surfaces—the inside and the outside. An examination will reveal that the Strip, having one continuous edge, produces *one* ring, twice the circumference of the original Strip with one half of a twist in it, which eventually *converges with itself.*

Since the middle of the last century, mathematicians have increasingly refused to accept a "solution" to a mathematical problem as "obviously true," for the "solution" often then becomes the problem. For example, it is certainly obvious that every piece of paper has two sides in the sense that an insect crawling on one side could not reach the other side without passing around an edge or boring a hole through the paper. Obvious—but false!

The Möbius Strip, in fact, presents only one mono-dimensional, continuous ring having no inside, no outside, no beginning, no end Converging with itself it symbolizes the structural kinship, the intimate relationship between subject and object, matter and energy, demonstrating the error of any attempt to bifurcate the observer and participant, the universe and man, into two or more systems of reality. All, all is unity.

I am indebted to Fay Zetlin, Artist-in-Residence at Old Dominion University in Virginia, who sensed the principle of convergence, of emergent transcendence, in the analogue of the Möbius Strip. This symbol may be said to crystallize my own continuing and expanding explorations into the unitary structure of all reality. Fay Zetlin's drawing of the Möbius Strip constitutes the visual image of this effort to emphasize the experience of coalescence.

R.N.A.

THE DOUBLE-EDGED HELIX
Science in the Real World

Preface

Recombinant DNA technology (often called gene-splicing), an offshoot of molecular biology aimed at the genetic manipulation of bacteria, plants, and animals, burst upon the scientific scene in 1973. For the first time, a relatively simple and general method for carrying out fundamental operations in genetic engineering at the molecular level became available. The primary concern of many scientists was, and still is, with the immediate uses and hazards of the new techniques. Comparatively little attention has been paid to possible long-range consequences of this technology, in spite of the fact that it is capable of producing novel organisms, not found in nature, whose behavior cannot fully be predicted. Coupled with other new genetic engineering techniques currently being developed, gene-splicing opens the door for human intervention to produce significant changes in the nature of life on earth. Some look for a panacea, but others fear that the current and immediately foreseeable applications of recombinant DNA technology, with their attendant risks, are only foreshadowings of powerful future applications that could threaten individual freedom, if they do not escape human control altogether. It is not surprising that a deep and sometimes emotional controversy over risks and benefits has developed between the more cautious and the less cautious factions in the science community.*

---

*The *scientific* community is made up of trained scientists actively engaged in producing new scientific knowledge. The *science* community includes not only the scientific community but all the associated administrative and ancillary activities of government, universities, institutions of various sorts, and commercial establishments.

This book was first conceived as a critique of the science community's attitudes toward recombinant DNA. But as I thought more about the controversy over this new technology, it became clear that there are numerous fundamental flaws in the entire scientific enterprise and its associated technologies. In many important instances science has become subservient to technology, which in our society is tuned to boundless growth and expansion. Ideally, science should be independent of this influence but not of human or societal needs. A real concern for these needs would require a serious evaluation of research efforts leading to long-range planning of the scientific enterprise and careful advance assessment of the applications of science, conditions that do not obtain at present. Unfortunately, preoccupation with the immediate goals of mission-oriented research, laudable as they may sound, has left little time for conscientious thought about alternate pathways for the use of science for humankind; social awareness has been seriously lacking.

A mature social conscience would demand an involvement of scientists in the politics of science, but not in any self-serving way. Leadership in the scientific community ought not to be of the *de facto* type that we now have; it ought to be planned, reasoned, and responsive to both scientists and the public with the leaders truly representative of the mass of scientists rather than of a select few. Today the scientific community as a whole is passively swept along in the wake of a relatively few science potentates whose wishes and whims set the fashions for the others. Most scientists, who in fact produce the bulk of knowledge, have little influence on the direction of inquiry, and little obligation is felt on the part of the *de facto* leaders to subject the premises or consequences of their chosen scientific directions to a rigorous analysis. In the end this behavior must be self-destructive for science. Unless these shortcomings are rectified, science cannot hope to remain viable and responsive to societal needs. If scientists wish to retain their credibility and their right to seek knowledge for its intrinsic value, an examination of end points, with public input, is going to be required.

The public's disenchantment with technology (which is

often equated with science), arising from the increasingly apparent negative side effects of technical innovations, has with the recent stringency in federal financial support for basic research, created a disquieting atmosphere in the scientific community, in sharp contrast to the science-society honeymoon of the 1950s and the post-Sputnik era. Scientists had come to expect limitless growth in public support with no strings attached. But in recent years, many scientists have been forced to adapt to a much more mission-oriented approach, and political activity has begun to play a predominant role in the whole enterprise, including biomedical research. Unfortunately, current research missions and political aims do not always represent a well-thought-out and widely supported policy, but are too often based on expediency. Political maneuvering has become abundantly evident at both national and local levels, as well as within institutes and universities. The somewhat arbitrary shifts and rearrangements have had profound and pervasive effects on scientific effort, producing low morale and a lower output of research. The political and economic realities have, however, sparked a maturing process. There are glimmers of enlightenment among some scientists, a gradual awakening to the shortcomings in the way science is being practiced with respect to human needs.

With the advent of recombinant DNA, molecular biology emerged from the realm of "pure" science. A discipline whose main purpose was to understand the fundamental characteristics of living systems was transformed, virtually overnight, into a force that not only can examine the living organism but now can manipulate the organism in ways never before possible, at the will of the scientist. For many biological scientists, recombinant DNA technology has brought the societal implications of biological research close to home for the first time. This in turn has raised questions about many aspects of the scientific enterprise as a whole; it has become clear that the advance of science and technology often exacts a high social price indeed. In the face of the many technologies already at our disposal, one is tempted to ask: Is it really necessary to add this new technology to the existing arsenal? Should not the commercial prospects of recombinant DNA technology be considered in the context of recent technological history, which does not have a very good track record in many areas? Should we, therefore, not proceed cautiously? Should we not try to put human concerns back into technology, and into science, its progenitor?

But why question the science community at a time when it seems to have given us virtually everything we have desired? Has not science provided us with innumerable great benefits? Consider: wonder drugs, plastics, pesticides, the transistor. . . . Science, in all its specialized fields, is a major force in our society, and through its handmaiden, technology, it has become the most significant single element in molding society. Science has provided the basis for an infinite variety of conveniences and services, and I do not suggest that we abandon useful discoveries or return to the vague "good old days." But we must ask ourselves whether a continuing procession of scientific discoveries and technological applications is what we need for the advancement of mankind. We already have an abundance of goods (whether or not they are equitably distributed), yet evidence abounds that we are experiencing a generalized malaise throughout the industrialized nations of the world, which strongly suggests that we do not need more hardware but that we should utilize more humanely what is already at hand.

This book discusses some aspects of the internal structure of the science community and is an account, as viewed by a scientist, of the relationship between science and societal issues. Most of my primary experience has been in the field of molecular biology, which forms a constant thread throughout the book, relating frequently to recombinant DNA. I have touched only briefly on technical material relating to the safety of recombinant DNA technology because this has been considered extensively elsewhere, particularly in the Environmental Impact Statement and the associated Guidelines for Recombinant DNA Research, issued by the National Institutes of Health in 1976. My concern has been primarily with the social and often philosophic components of science and science-based technology that inevitably intersect with economic and political forces.

The subjects discussed in each chapter, though related to one another, are treated as discrete entities; each chapter is complete in itself and no fixed order of reading is necessary. To maintain the continuity of the underlying theme I have referred to or discussed some material more than once, but this repetition is minimal. The first part of the chapter on gene-splicing is scientific. The approach, although unorthodox and somewhat simplified, is conceptually sound. It was designed to provide the lay reader with an intuitive grasp of molecular biology and it is not intended as a primer.

This book has been written in the context of the emergence of science as a uniquely influential enterprise in the middle of the twentieth century. It is an attempt to show that science could help, albeit in a way to which it is not accustomed, solve some of our societal problems. The approach is general and offers no specific recipes. Furthermore, I cannot and do not make the assumption that appropriate decisions regarding the utilization of recombinant DNA technology will eliminate the ills resulting from other technologies; but conscientious thought about one set of problems can serve as a model and will be instructive for all of us. I hope that the vantage point of a scientist who has actively practiced during this period of explosive growth of biomedical science will provide a unique perspective and, in the words of the social philosopher Jacques Ellul, will awaken the reader.

1
Dilemmas

We live in a time of literal surfeit of products and processes conceived and generated by contemporary science. This is not science purely as knowledge, confined to the ivory tower, but science that has escaped from it; for the scientific result that remains in the laboratory can have no direct societal impact. What we are considering, then, is science and the scientific way of thinking that have been translated into technological innovation.

Technological subvention, although often useful, has not been entirely successful in our time, in spite of antibiotics and moon landings. More and more frequently we are faced with problems created by "solutions" to the problems arising from previous technologies. For example, increasing food production by the use of more fertilizer leads to water pollution, but the procedure seems natural and necessary to the managers of society. What we are witnessing in general are the results of a technocratic system that lacks a self-correcting servomechanism; there are no built-in provisions for monitoring new developments in different fields and adjusting them in order to optimize the well-being of society as a whole. The system is characterized by a contradiction in its *modus operandi*, leading inevitably to a dilemma: on the one hand, the system inspires an unquenchable desire by society for "progress"; on the other hand, it often delivers the goods at an unacceptably high social cost, which includes a price paid not only for services rendered but also for consequent, undesired side effects. Why do we continue to follow such an irrational course, one that has led to severe societal problems?

The reason is cyclic. Science creates, technology applies, and applications create problems; these problems are then thrown into the science-technology hopper for solutions. But the modern science establishment, including a monumental administrative component in government and private organizations as well as practicing scientists and engineers, has become a sprawling giant with its own momentum, admirably capable of solving technical problems but not oriented toward the consideration of societal needs in the broad context. Science is, however, of necessity committed to its sources of support: government (including the military) and industry. They themselves are inextricably intertwined to form what some call the corporate state, the single most important determinant of modern industrialized society, characterized by a primary drive for self-perpetuation and expansion. The corporate state controls the economy, and in so doing it mandates, directly or indirectly, the direction and growth of science and technology. Economic necessity thus presses the public to accept indiscriminately the technological system as a whole, in spite of its antisocial tendencies. Society is continually presented with a classical dilemma—a choice among alternatives, each of which has adverse features. Somehow I think we have outdone ourselves in our willingness to accept so much that is bad for so little that is good.

Science, the ur-source of the industrial commodities that form the basis of the economy, has thus become an affair of state, and the pursuit of science has become a politically and ethically charged activity, whether or not we wish it to be so. In a recent discussion, André Cournand pointed out the need to develop an ethic of science influenced by the impact of science on society.[1] He noted that the applications of science have accelerated a kind of sociocultural evolution, which he aptly called "blind emergence." In a general way he suggested that the evils of uncontrolled development be identified and discussed by scientists on a worldwide basis, creating, in effect, a universal scientific community. This concept has also been put forward by D. Dubarle, who wrote in 1963: "The calling of the scientific community requires it to escape from its context, to shake off the controlling hand of the particular governments and to set up its own worldwide institutions, other than those in which the different national states can see themselves reflected and recognize their national soul still held down by their attachments to the soil. . . . The turn taken by our history has made it clear that the present system of scientific activity cannot survive for very long."[2]

In a somewhat similar vein, Jacob Bronowski has proposed the "disestablishment" of science; he suggests that national funds be set aside and distributed in ways involving no commitments on the part of scientists.[3] Jacques Monod[4] has taken the extreme position that objective knowledge is supreme, that it is superior to man himself—a view that modern society is unlikely to find acceptable. Although some of these approaches have attractive features, they all smack of scientific elitism. What is called for instead is a demystification of science, which together with a reevaluation of societal goals could provide a basis for better judgment on technological issues by all who are exposed to their consequences.

It is unfortunate that, individually and collectively, we have been conditioned to accept the philosophic view that technology (and hence science) is the great healer, that its purpose is to find solutions. This feeling is more ingrained than we would like to admit. How often have we traversed the path from luxury to convenience to necessity? Our needs have been modulated and expanded, often hedonistically, in order to feed the growth of the technological system. Mass advertising has been used to submerge and camouflage the negative aspects of technology and create the illusion that we can have it both ways— endless benefits with negligible cost or risk. The technological dilemmas have been masked.

Scientists, who are also subject to this conditioning, nonetheless bear a special responsibility toward the public to take an objective view of their domain and serve as guides in separating the scientific and technological "solutions" that are truly life-enhancing from those that are, on balance, irrelevant or detrimental; a responsibility to recognize not only the intrinsic values of science, which indeed exist, but also its limitations in ameliorating the human condition. I am aware that this demand poses a special dilemma for the scientific community, for in spite of the high-powered and sophisticated research potential of Big Science, the scientific community is not entirely free to choose a rational path whose goal is the satisfaction of true needs. Moreover, the scientist himself is presented with a dilemma. The scale of Big Science tends to make research impersonal; the individual scientist is discouraged from thinking about his work in a broad social context. There has as yet been no widespread movement on the part of scientists to identify and evaluate the aims and societal implications of their work, or to steer it deliberately in altruistic directions. To a large and increasing extent, economic considerations outside the control of science determine its direction today.

How are we to escape from these dilemmas and break out of the cycle of irrational scientific and technological growth? I cannot offer a recipe here, but an essential ingredient is surely a mature social consciousness on the part of scientists as well as the general public. For the first time, the scientist is called upon in the gravest tones to insert subjectivity, not into his method, but into the results of his work; he has to become a judge.

2
A Scientist Looks at Science: An Overview

The science of DNA[1] was born, for all practical purposes, in the 1940s. It was a particularly exhilarating period for those of us who began our careers at that time; we saw biomedical science emerge from an amorphous mass of apparently disjointed facts and become a highly ordered edifice. Of course, pockets here and there of biochemical knowledge had provided a limited scientific framework for biology, but in the main a broad overview was lacking. With DNA, bioscience took on a new look. We began to seek wider horizons, for it seemed clear from the start that DNA was to play a major part in future biological thought. This new brand of science catalyzed its own growth. Each new finding seemed to beckon others, pressing forward and fanning out in every imaginable direction: DNA touched virtually everything in biology.

We were present when new theories were born—and died. We saw major breakthroughs as well as the slow, persistent accumulation of data that is the keystone of experimental science: we felt the growth pains. Science entered a golden period just after World War II,[2] when it changed from Little to Big; to a time when obscure professors were raised to important posts in government and influential positions in the community;[3] from a time when a researcher felt free to solve a small scientific puzzle without worrying whether the solution was publishable to a time of publish or perish; from a time when scientific discourse was relatively free to a time when the diameter of the scientific circle diminished and finally collapsed into a handful of cliques; from a time when the pursuit of truth was the primary aim of science to an era of "relevant" science; from a time when the investigator built or repaired his own instruments to a time when he became helplessly dependent upon highly trained electronics engineers to repair a high-speed circuit in a computer or the power supply of a laser beam.

Science can be exhilarating—when a scientist's prediction based on past research is confirmed by experiment; it can be excruciatingly painful—when a carefully developed and cherished hypothesis is demolished. Science is usually intense, always requires painstaking effort, and yields results begrudgingly. Advances occur slowly. Major discoveries are infrequent. Often an investigator's results must await subsequent experiments by other scientists before a meaningful concept can emerge. When a new finding seems to be a breakthrough, throngs of scientists join the search for new knowledge. This has happened a number of times in biological research during the past three decades: new paradigms have been created. In 1944 it was shown that DNA is the genetic substance,[4] but the force of this discovery was not felt until 1953 when Watson and Crick announced their hypothesis for the structure and function of DNA.[5] This opened up one of the most important paradigms in modern biology. Following this discovery a new mood of excitement arose in laboratories around the world; an aura of universality prevailed, for DNA was recognized as an eternal and deep truth.[6] There was an immediate increase in the intensity of biological research; the number of publications skyrocketed, and fervor mounted at scientific conferences where DNA was being discussed.[7] Over the next decade the structure of DNA gave birth to many other concepts in molecular biology, each one accompanied by a new burst of enthusiasm.

But not all laboratories were at the forefront of the excitement. During this period quiet, inconspicuous research was being conducted on an unexplained feature of the infection of bacteria by certain viruses called bacteriophages: the surprising fact that the bacteria sometimes destroyed the infecting virus, rather than vice versa. This line of research was only one of many sophisticated molecular genetic approaches to fundamental questions in biology. At that time no one had the slightest inkling that this research would one day rock the scientific world. The work continued unobtrusively for a number of years; during this time, a battery of solid genetic data from bacteria and bacterial viruses was amassed.[8]

But this research was not to remain quiet for long. Interest increased when the research predicted the existence of restriction enzymes, agents that seemed to protect the bacteria by killing the virus. Although the first restriction enzyme was isolated and characterized by H. O. Smith and co-workers in 1970, it was not until these enzymes were isolated from E. coli in 1973 that all hell broke loose.[9] The restriction enzymes, which can inactivate viruses by making specific cuts in their DNA, made it possible for the first time to produce specific DNA fragments and to recombine segments of DNA from different sources, paving the way for recombinant DNA technology. The method was simple, rapid, and precise. It was now possible to rearrange the natural genetic elements of distant species, producing genetic determinants not seen before in nature: hybrid molecules of DNA. Another paradigm had been created; new biological domains, not heretofore readily accessible, were open to investigation. For example, the study of the structure and function of mammalian genes was greatly facilitated, for functional genetic subunits could be separated, reassorted, and studied individually. The synthesis of drugs and foreign proteins in bacteria was made possible; genetic engineering of higher organisms by the use of recombinant DNA techniques became an enticing prospect for some scientists. The overwhelming potential of recombinant DNA technology was apparent from the outset. The excitement spread like wildfire throughout the science community, rekindling dying embers in some laboratories and fanning flames in others, both large and small. And the fires still burn.

The atmosphere generated by the advent of recombinant DNA technology is not unique in the history of recent biomedical research. Periodic breakthroughs have regularly peaked the already high level of activity, giving way on some occasions to a frenzied atmosphere. For example, in the early 1960s, basic discoveries by Marshall Nirenberg, H. G. Khorana, and Robert Holley paved the way for the cracking of the genetic code by providing experimental evidence that genetic information in DNA comes in groups of three. A "word" in a DNA molecule was shown to consist of three chemical groups, or "letters"; and the meaning of the "word" depends on the order of the groups.[10] The Nobel Prize was up for grabs. Research on the genetic code immediately became an area of hot pursuit; hastily written and often incomplete papers were common, and there was partisanship and overt rivalry among scientists. Nirenberg, Khorana, and Holley eventually beat their opponents in the race for the Nobel Prize. Again, work in an exciting area of hormone research also produced an extremely tense and bitter atmosphere, which lasted for more than twenty years in a number of laboratories.[11] The principal opponents in this race were Andrew Schally and Roger Guillemin. Both men were studying the structure of very elusive hormones present in animal tissues in such small amounts that horrendous technical difficulties were encountered. This time, the Nobel Prize was finally shared by the opponents.

While these episodes—like their well-known prototype, the rivalry of Watson and Crick with Linus Pauling, described in *The Double Helix*[12]—make exciting copy, the entire exercise of competitiveness in many ways serves to undermine scientific effort and gives rise to distorted values in the minds of scientists. Accolades and peer approval are part of the game, to be sure. But the disinterested pursuit of truth does not always hold up well in the face of these extraneous pressures.[13] I do not mean to imply dishonesty, nor do I suggest that competition is new in science—hardly so. But along with the increased intensity of all activities associated with science these days, including fund-raising, certainly the magnitude of competitiveness has increased. Under this pressure the academic atmosphere comes to assume an aspect of intense technology. Competitive pressures also give rise to secrecy.[14] Another factor that tends to decrease the quality of research relates to the "publish or perish" syndrome, which is now worse than ever. Prodded by a decrease in federal funding, this pressure has crept insidiously into the investigator's daily life. The scientist is apt to be less rigorous when exploring the peculiarities in his results; he may rationalize any uncertainties, reasoning that another paper will look good in his bibliography. Then there is the question of frankly fragmentary or preliminary results that find their way into print prematurely, often side-tracking other scientists who try in vain to reproduce the reported work.

The frenzied competitive situation was exacerbated in the late 1960s when funds for basic research were seriously decreased. Grant applications became the bane of the scientist's existence, and "grantsmanship" (a euphemism for the ability to "dress up" an application) became a prime asset. There is nothing wrong with presenting a polished, well-documented application, but when the main focus is on how to win a grant, science takes second place and is bound to suffer.

The peer-review system of evaluating grant applications is generally used in this country to determine the distribution of funds. Reviewers are practicing scientists who work in the applicant's field or a related one. When the supply of funds shrinks, however, only the most highly ranked applications they approve can be funded. Many "approved but not funded" applications represent high-quality research that in the past would have been successful; in many cases the decision is quite arbitrary. The unfunded researcher is often justified in feeling unfairly rejected. Moreover, when the unsuccessful applicant himself next sits as judge on a peer-review panel, he will be more inclined to be rigorous in his evaluations—perhaps too rigorous, contributing still more to an aggravated situation.[15] Ultimately the fault lies with the federal government, which in the 1950s created and encouraged a vast scientific enterprise that could not be maintained.

Associated with the decrease in funding has been a shift in the nature of the research programs chosen for support. "Safe" projects, targeted studies in circumscribed areas designed to support a popular assumption or settle an already half solved question, tend to find favor. This policy discourages innovative and original ideas, which might be considered "chancy."[16] By the 1970s we had entered an era of pragmatic science, a hard pill to swallow for those of us who had become accustomed to the ivory tower. The practice of science had changed in a qualitative way, although the intent of the funding agencies was meant to be only quantitative.[17]

For the individual investigator the course is clear: he goes where the money is; he has no other choice. Frequently this means a change of direction: the study of a new literature, the design of different kinds of experiments, perhaps new instrumentation, and in general a different point of view. Coupled with this is the increased urgency to publish in the new area of research. The scientist has got to sing for his supper, and sing to a changing tune of uncertain authorship and unclear significance in the overall scheme of things.

In America, biomedical science policy has consisted largely of a series of responses to spurts of enthusiasm in Congress and uninformed political pressure for new "programs." Policy decisions have been based more on temporary expediency than on a hard look into the future; and the implicit guidelines are, in general, economic ones common to all aspects of the technocratic system. No concerted attempt has been made to integrate biomedical policy with other social policies. For example, there is now new emphasis on research on the causes of aging; this may be a laudable topic, but the effort is not correlated with economic and social policies aimed at easing the burdens of the aged. Another major recent policy decision was embodied in the National Cancer Act of 1971, which has recently come under fire because scientists have not delivered the hoped-for cure. In 1971 many scientists undoubtedly voiced optimism about potential cures, but I doubt that anyone promised a firm delivery date. However, no one disillusioned Congress either, and so a large program was initiated. The program was bound to be a disappointment because the basic science that may be applicable to cancer problems was not fully developed. As has become customary in the era of Big Science, the scientific establishment felt it necessary to be overly enthusiastic to a Congress attuned to quick fixes.

Now, the advent of recombinant DNA technology with its

far-reaching implications has, with government encourage-
ment, transformed biological science (much as atomic fission
transformed physics) to a powerful technological and economic force that cannot be relied on to remain benign. But, as in
the past, little thought is being given to long-range consequences. (In chapter 4 I examine the implications of the new
technology and indicate how it may come to function in a
technological society).

As one who has lived in the biomedical scientific community for a number of years, enjoying freedom of inquiry, federal
support, and a sense of fulfillment, I may seem ungrateful
when I criticize the establishment that has provided these
privileges. But my remarks are not meant to be destructive—
quite the opposite. The scientific enterprise plays such a central role in modern society that a serious reevaluation of its
goals has been in order for some time.

Traditionally, scientists have held themselves accountable
only to their peers, who saw to it that high standards of work
were maintained. An aloofness from the public has resulted,
and understandably so since the general public could not be
expected to participate in a discipline that requires years of
specialized training; herein lay the seeds of elitism. The increasing sophistication of science has increased that elitism.
Some years ago, in a discussion of the scientist's social responsibility and the applications of science, the Nobel laureates
Percy Bridgeman, I. I. Rabi, and Ernst Chain expressed the
prevailing attitude. Chain said: "Let me first of all state that
science, as long as it limits itself to the descriptive study of the
laws of Nature, has no moral or ethical quality, and this applies to the physical as well as the biological sciences."
Bridgeman said: "From the point of view of society, the justification for the favored position of the scientist is that the scientist cannot make his contribution unless he is free, and that
the value of his contribution is worth the price society pays for
it." Rabi added: ". . . the scientist cannot take the responsibility for the manner in which society utilizes the knowledge he
uncovers.."[18] There can be no doubt that these scientists were
conscientious, thoughtful people; but they were surprisingly

innocent. Unfortunately, their words carry a great deal of weight—they are Nobel laureates—but their scientific expertise should not automatically qualify them as savants in other areas. The fact is that these men, like so many other scientists, take a limited view of the implications of science, reflecting an underdeveloped social conscience. As Harold C. Urey, also a Nobel laureate, pointed out, in response to these attitudes: "We do not hold the miner responsible for the use of the iron which he mines from the earth, but it would be quite right and proper for him as a citizen to object to its being shipped to Japan as scrap iron to be used in a war against his country at some future date."[18]

Modern scientific research in this country has become increasingly mission-oriented. When the research is economically important, its results are often applied with little loss of time; such was the situation with the transistor, the heart of modern electronics. In that case, and in many others, science acted as an arm of technology. It is foolish for scientists to close their eyes to this reality, when they should be guarding science against abuse and exploitation for commercial purposes that have little to do with either human needs or the acquisition of pure knowledge. In traditional fashion, as I discuss in later chapters, most scientists have not felt the need to become involved in the application of their discoveries; indeed, they have carefully avoided any such intervention, arguing that this is not their domain. This simplistic notion, which came into vogue about 150 years ago, is irrelevant and even dangerous in modern times. In the face of recombinant DNA technology, which will most certainly affect the lives of future humans, this archaic view of the pursuit of knowledge is especially in need of substantial updating, for if the scientific community will not guard the public interest when a powerful but highly esoteric new scientific technique is discovered, who will?

Contrary to prevailing fears, the acceptance of public accountability and responsibility by the scientific community would not preclude the pursuit of knowledge for its own sake. In fact, as I have pointed out, that pursuit is currently being phased out with no good justification.

Recombinant DNA technology has immense societal implications, embodying applications to medicine, agriculture, and industry; its possible influence on ecological systems and future generations of humans is incalculable. It will permit manipulation of the gene pool of the earth, and thus manipulation of the nature of all life. At this time, techniques have already been developed by which genes, which are composed of DNA, can be shuffled about so that DNA from any source —say, animals or viruses or fruit flies—can be inserted into living bacteria. Inside the bacteria these genes can be made to perform their normal functions, if all goes according to Hoyle, even though they are outside their normal habitat. I discuss this aspect of recombinant DNA technology in later chapters; it is sufficient to say here that recombinant DNA presents scientists with a new and uniquely powerful means for altering living cells according to their design. One Nobel laureate has said: "We can outdo evolution."[19] The biological scientists' responsibility is therefore immense; it is as great, or greater, than that which fell upon physicists a few decades ago. I suspect that many of the implications of this technology have been cast aside by the scientific community because a more enlightened view would require a general examination of societal problems, and the solutions to those problems might place constraints on the scientific enterprise.

Meanwhile, although the public awe of science continues, there is a growing uneasiness about technology.[20] Someday, as the nuclear, ecological, and now genetic hazards and threats grow larger, this unease is likely to erupt with destructive force as a full-scale antiscientific and antiintellectual movement.

How did science come to occupy its unique position of high-regard and virtually zero accountability? After the somewhat constraining views of Aristotelian-Ptolomaeic science, Western civilization viewed the scientific revolution that began in the seventeenth century as a breath of fresh air, an opportunity to reexamine and expand the bounds of inquiry that had been imposed on human thought and had placed limits on action and experimentation. Ancient science observed the whole; it perceived relationships without manipulating the corpus or breaking it down into its component parts. Reductionism was forbidden, not by edict, but by the culture of the times, which was a more efficient and powerful mechanism than law. An about-face took place when modern science appeared on the horizon. The Copernican hypothesis and its later experimental verfication by Galileo, who showed that the earth revolves around the sun rather than vice versa, did much to root out simplistic concepts of the structure of the universe and its relation to God. The early successes of experimentation dealt a heavy blow to the religious philosophy of the time. As a grip of religion weakened, the stage was set for the cultural acceptance of a new world view, based only on verifiable and tangible facts. The religious upheaval created a void that the new science, developed in its modern idiom by Bacon and Descartes, was ideally suited to fill. The logic of the method, both theoretical and experimental, produced results that inevitably led to universal acceptance: for the doubter, proof could be furnished. The position of science was thus solidified early, particularly by the work of Isaac Newton. Moreover, because the need for discipline is intrinsic to the scientific method, the autonomy and self-regulation of science seemed reasonable; public accountability was unnecessary. Accountability within the scientific community was a function assumed by the academies, the guardians of the methodologic ethic. The public, unable to appreciate the substance of science, and only distantly and indirectly affected by scientific development, was not involved.

Scientists as a group, notably those in France and Germany, have enjoyed respect and prestige since the end of the eighteenth century. When the impact of science on society became palpable, science, with its precision and instrúmentation, began to be recognized as a prime source of hard facts about the natural world. Other fields waned in importance. As the amateurs of science left the scene, during the latter part of the eighteenth century, the language and concepts of science became more esoteric, creating the false impression that science was a domain for geniuses, not for ordinary people. This attitude set the stage for the separation of scientists from ordinary society.

Whereas science in Europe was placed in a unique position,

the same was not true in America. Americans had learned early on to idolize technology rather than science. The support of science per se in America was minimal before the Second World War; science was a minor activity carried on in small, not too well-equipped university laboratories. The transition from Little to Big Science, which began after World War II, was spurred by important wartime developments like antibiotics and plastics, and by the leftover machinery of wartime research, which provided a physical as well as psychological basis for the continuation of large-scale efforts especially in physics and engineering. Science was ready to move forward.

Developments realized during the war, such as atomic fission and radar, provided a broad substratum on which to build and expand the scientific effort. An immediate post-World War II problem concerned possible uses for the new-found atomic energy. Special Senate hearings brought politicians and scientists into intimate contact for the first time. The *Congressional Record* of that period makes interesting reading; in its pages scientists were held in awe and even revered.[21] For example, Senator Tydings said: "There are a few men . . . or maybe several thousand in the world whose mental development in many lines—and particularly in the scientific line—is like comparing a mountain to a molehill when you compare them to the rest of us." Senator Russell remarked: "My attitude toward scientists is . . . pretty much like the boy living in the country and going to the country doctor. He thinks the doctor can do anything."

It was not long before physicists and engineers were herded into government offices and asked for all sorts of advice regarding nuclear energy. Men like Vannevar Bush, who had headed the Office of Research and Development during the war, were only too happy to comply. After all, scientists and stodgy professors had had a lean and unappreciated existence for a long time. Bush expended a great deal of effort to find a place in the sun for science and scientists. Having been in close contact with the development of atomic energy, he had lavish visions of its application to peacetime uses such as the generation of power; he was excited about operations research and systems analysis. He helped create the National Science Foundation. In a book entitled *Endless Horizons* Bush outlined a scheme for the limitless enterprise of science.[22] The fever was catching. Once having sat in the majestic and impressive Washington offices, scientific advisers now created their own momentum. They began to push science, and their effort was aided from other quarters. It is difficult to know which was the primary catalyst—government or the private citizenry.[23] In any event, science had entered the big time, trailing dreams and panaceas.

The global thinking in Washington, with its attendant politics, soon oozed out into the general scientific community. During the late 1950s and early '60s the biomedical community was literally plied with federal funds. From a scattered and uncoordinated pursuit, biomedical research crystalized into a large and cohesive enterprise, now part of the government establishment. A symbiotic relationship was established, as scientists offered advice to the government on biomedical problems and health care. And the government listened—for awhile.

By the late 1960s, members of Congress began to respond to the disillusionment of constituents conditioned by a number of recessions, increasing inflation, and the increasingly apparent ills of technology. Research appropriations were cut. In spite of outcries from a now overblown science community, there was no real relenting on the decision. The golden era had come and gone. Scientists now had to account for their activities as they never had done before. The Johnson administration and its successors emphasized practical applications, and Congress followed the lead. But the orientation and resources of the science community, molded by past excesses, could not so quickly be brought into line with political expectation. Dissatisfaction became evident on both sides.

A review of the biomedical research effort in America was consequently undertaken in 1975. The President's Biomedical Research Panel,[24] composed of about 150 distinguished scientists from all sectors of the biomedical community, was appointed to assess the conduct, support, policies, and management of biomedical and behavioral research supported by the National Institutes of Health. This episode is instructive because it reveals a deeply ingrained attitude on the part of the scientific community toward social responsibility and the conduct of science.

The President's panel carried out an extensive and scholarly study. It met 30 times in 15 months, questioned 160 witnesses, and received written testimony from 277 health-related organizations. It submitted its report in April 1976. When the report was discussed before the Senate Subcommittee on Health in June of that year, it immediately became obvious that scientists and senators were at cross-purposes. The subcommittee questioned the panelists in five areas, none of which was covered in the report, and then the report was shelved—much to the chagrin of the panelists.[25] The reason for this apparent snub to the scientific community was pointed out later by a congressional science fellow involved in the process, Dr. A. M. Silverstein.[26] According to Silverstein, the subcommittee had informed the panel that the real purpose of the study was to analyze the relationship of biomedical research to societal problems. Silverstein noted the issues that "interested Chairman Kennedy (e.g., health technology assessment and transfer; the moral aspects of and societal involvement in biomedical research decisions; cost-benefit aspects of basic research in the context of an increasingly expensive health care system, etc.)." Instead, the panel took a narrow view of scientific interests, gave a clean bill of health to the National Institutes of Health and biomedical research in general, and paid little attention to the broader questions of interest to the senators.[27] Panel members chose to remain aloof, not to say elitist, in a political process of which they were a part. They said, in effect, that science should be judged by its own standards. In this blind attempt to maintain the status of the past, which is widespread even among scientists who consider themselves socially progressive, the panel ignored the political realities of Big Science and chose not to participate meaningfully in the process that will eventually decide the fate of science. Instead of jumping at the opportunity to add scientific input to societal issues, the scientific community is prone to regard any coupling of scientific and social issues as antiscience. In the end, this is the best way of ensuring that much that scientists prize will become vestigial.

During the past decade, "antiscience" movements have been of concern to scientists and have been discussed at a number of important symposia.[28] That these movements are primarily antitechnology rather than antiscience has not been taken to heart. The president of the National Academy of Sciences commented, in 1970, that part of the trouble had been caused "by scientists who exaggerate the all-too-genuine deterioration of the environment." He then expressed the view: "I much prefer that we attempt to manage our technological civilization yet more successfully, remedying the errors of the past, building the glorious world that only science-based technology can make possible."[29] The high priest thus rejected out of hand any departure from established procedures. The same viewpoint, in less blatant form, is widespread. In a scholarly study of the state of American academic science, Smith and Karlesky examined a variety of factors that play an important part in modern research.[30] Throughout their book it is clear that scientists and university administrators are concerned about the health of their endeavor, which is certainly proper. But I was struck by the fact that in their minds the "health" of American science did not have a component in it directly related to societal needs. One is forced to recognize the implicit assumption of scientists and their mentors that science is naturally good for society; that scientists need not explicitly spell out what can be done, since Science will assuredly solve society's problems without making any special effort in that direction. This attitude reminds one of C. F. Wilson's famous statement to the effect that "What's good for General Motors is good for the country." In this complacent philosophy, coupled with the directing influence of federalized funding on science, one can perceive the seeds of the technological fix and its influence in shaping the social consciousness of scientists and the public.

The social consciousness of most scientists does not extend to their own sphere of activities. This is not a criticism; it is an observation. For example, many molecular biologists take pride in their "liberal" political views: they marched in pro-

test of the Southeast Asia war; they fought against the use of
chemical and biological warfare; they decry radioactive contamination by nuclear wastes; they abhor pollution.[31] In brief,
their values seem to be related to the bettering of the human
condition. Yet in their own realm many of the same scientists
fail to take note of the possible ill effects that could follow from
their work; they make the implicit, vague assumption that all
science is good, as though its beneficent application were foolproof. This leads to the illogical conclusion that any and all
goals are equally desirable in the search for knowledge, and
this is somehow connected with freedom of inquiry. Scientists
are rightly concerned about freedom of inquiry. But when it is
discussed, insistence upon the neutrality of science often
aborts rational analysis. Some scientists hold up the specters
of Galileo or Lysenko at any suggestion of public accountability, although their histories are not relevant to the issues of
public and environmental safety raised, for example, by recombinant DNA technology.[32] Scientists still feel comfortable
with seventeenth-century arguments concerning knowledge
and truth, arguments that take no account of modern technological society and the accelerated impact of science on everyone.

A common feature of technologies is that they respond first
to the needs of the industrial structure that spawned them and
second, if these do not interfere with the first, to human needs.
This is the immutable contradiction of our industrial system;
it is a system that, by design, depends on production and
growth. The physical realities of finite energy supplies, the
limited ability of the environment to absorb pollution, population growth and the finite potential for food production, and,
ultimately, the projected thermal instability of the planet force
the inevitable conclusion that growth must cease within a few
decades. By anyone's calculus there can be no setting aside of
this dilemma. The choice is clear: let matters proceed in a
more or less random fashion to the natural and ominous end
point; or try to transform the present socioeconomic structure
based on unrestrained technology, by developing appropriate
controls. The practice of science as we know it cannot contin-

ue unrestrained, in the present milieu, for its results are bound to be applied by the industrial establishment in the name of progress. But, as the scientist Bentley Glass asks, ". . . can we honestly set aside the conclusion that *progress,* in the sense of ever-growing power over the environment, must soon come to an end?"[33]

In attempts to maximize the best and minimize the worst, technocrats place a high degree of confidence in cost-benefit (or risk-benefit) analysis. But such analysis becomes more irrelevant as time goes on; indeed in many areas, such as the alteration of the landscape for industrial reasons, cost-benefit analysis is completely inadequate. Asthetic, ethical, and moral questions involve value judgments, to which the "hard" numbers required for cost-benefit analysis cannot be assigned. Decisions involving those questions must therefore be political, not technical.

It is not so much technology, itself, as its present vast scale, that creates the problems. The application of science to the development of intermediate and alternative technologies could be highly beneficial.[34] Such technologies emphasize natural processes, the use of renewable resources, labor-intensive instead of energy-intensive production, and minimal waste. Intermediate technology does not call for renunciation of scientific principles or a return to the untamed wild; quite the contrary. For example, a group of pioneering young scientists called the New Alchemists[35] uses the most advanced scientific knowledge to achieve nearly self-sustaining family-sized units for food production. Giant technological approaches are in general renounced. Philosophically this approach is capable of achieving a state of human fulfillment not possible with a surfeit of material goods. We have already proved that a plethora of hardware, drugs, and consumer goods have not achieved this aim. The convergence of so many technologies has reduced the public to a listless, frustrated mass of humanity without a meaningful function. Another technology will not solve the problem; that will require a monumental and courageous political decision, backed by the determination of all of us.

In the face of such fundamental problems, cries by scientists for freedom of inquiry seem banal, self-serving, and irrelevant. The cries are a result of what Theodore Roszak[36] has called the "single vision" of science—the view that the content of human life can be comprehended only through a scientific understanding of its inner machinery, by a complete dissection and analysis. This reductionist philosophy has created and nurtured the technological state, and it has done so at the expense of the value of wholeness. This is an unfortunate outcome for science, which does not inevitably demand application as technology; nor is science incompatible with other, more humane philosophies. It is the emotional and commercial content of science, put there by our culture, that has led directly to the problems. Science practiced in a newly responsible way could play a vital role in extricating society from the impending crisis. But this means that scientists will have to develop a social conscience, convey this to the people, and above all, teach their newly acquired wisdom to the technocrats.

To call for an awakening of scientists, technocrats, and the masses on whom technology is practiced sounds all but hopeless, to be sure. But there is no other way to halt the impending technological disaster. Scientists have had freedom from accountability and responsibility for a very long time. They have the knowledge and the qualifications necessary to recognize the dangers of our present technological course, and they cannot escape from the moral responsibility of acting to change it—even at the sacrifice of cherished prerogatives.

3
Gene-Splicing

In spite of modern electronic expertise, so evident, for example, in weaponry and space exploration, the most sophisticated manmade machines cannot compete in the performance of complex operations with the simplest organisms found in nature. The distinguishing characteristic that sets living organisms apart from complex systems devised by man is the capacity of the former to maintain normal internal stability in the face of external change. This is so whether we consider single cells or a large ensemble of cells. For example, a human who is exposed to rather wide extremes of external temperature still maintains a constant bodily temperature. This stability, called *homeostasis,* is seen not only with temperature but with many other physiological characteristics. True homeostasis is the result of a large number of coordinate control systems. The essence of these systems is that they function at the level of molecules, which means that information is transmitted by molecular interactions that take place in infinitely small spaces. Manmade machines are not apt to achieve this ultimate aim—minimal space. With recombinant DNA (gene-splicing) at his disposal, man has at least the possibility of altering (if not of creating) a molecular control system according to design.

Control mechanisms in biological systems are responsible for the systems' survival; without these controls, environmen-

41

tal changes would often lead to death. Numerous control elements exist in natural systems; their number and complexity depend on the function controlled, which in turn is determined by the kind of cell. Control systems operate through feedback, positive or negative, which means that the product or end result serves as input. A room thermostat provides negative feedback: to prevent overheating, the sensor, responding to the input—heat—turns off the source of heat. In biological systems that yield chemical products, the product may control its own production; for example, when the amount of the product in a cell rises above a certain value, it may then inactivate one of the components used in its synthesis.[1]

Bacterial cells are one thousandth the size of animal cells, and are much simpler; consequently, they require a less elaborate system of controls. Many bacterial control systems are now understood, but very few have been analyzed in animal cells. Biological control systems, as I indicated earlier, operate at the molecular level. In studying them, scientists are therefore concerned with the nature of chemical reactions that occur among molecules in the cell, in particular among large master molecules such as proteins and nucleic acids, which control the synthesis of other molecules. A major goal of recombinant DNA technology is to study the structure of animal DNA with the aim of learning how the various control systems function when a specific piece of information in the DNA is selected for translation into a protein molecule needed by the cell. How do the organism's needs call forth from its DNA storehouse the required response and no other?

The study of chemical interactions in biological systems is in reality a study of electrical, or more precisely, electronic interactions among the various cellular components; in fact, all chemical interactions (or reactions) occur through electronic mechanisms. In principle, one can achieve a basic understanding of molecular biology if the following view is adopted: Reacting molecular species are simply minute masses of matter—as small as one can imagine—on whose surfaces exist specific configurations of charges. The charges, which may be positive or negative, similar to those of the common bar magnet, cause attraction or repulsion between molecules that have unlike or like charges.

The molecules most frequently studied in molecular biological systems are composed of the chemical elements carbon, hydrogen, oxygen, nitrogen, sulfur, and phosphorus. These make up the minute masses of the molecules. Specific arrangements of these elements determine specific surfaces in the resulting molecules; a particular spatial arrangement of chemical elements is always accompanied by a specific electronic configuration, which has a definite electrical charge. A large molecule possesses many charges and may be strongly attracted or repelled by another large molecule. Molecular biology is a study of the interactions of large molecules.

The secret of these interactions lies in the principle of complementarity, as in a lock-and-key mechanism; the large number of charges and their exact positioning provides the specificity so crucial to biological systems. This specificity assures that biological systems behave in a fixed manner and are not frequently capricious. Complementarity requires two features: the surfaces of the two molecules must fit together, as a key fits into a lock, and the charges on these surfaces must be opposite, so that electrical attraction is present. If only one feature exists, the interaction is transitory and there is no permanent outcome; molecules simply move away from each other to seek other more suitable partners with which reactions can occur.

In biological systems, a successful interaction between two molecules is only the initial step in a whole series of reactions. In general each reaction will yield a product, a new chemical, which will in turn be used for another purpose in a different reaction. This happens repeatedly, creating a network of interrelated reactions.[2] Usually there is one final product, a protein, which is designed to perform a specific function. Thus an important function of pancreatic cells is to produce the protein insulin, but the cell could not make insulin unless all intermediate products were also made. Although these intermediate products are essential for the cell's function, by themselves they would not be able to control the level of sugar in the blood, as insulin does.

The cell is a highly ordered system; its reactions are coordinated. Order is maintained through control elements that are themselves molecules or parts of molecules. Consider an important problem that occupies molecular biologists much of the time: the flow of genetic information from DNA, which eventually leads to the production of all protein molecules in the cell.

DNA is a long molecule that contains the information for not just one but many proteins; it contains many genes, and each gene provides the information for a protein. A gene is nothing more than a specific series of chemical groups composed of chemical elements; each group is characterized by a mass-charge configuration. There are four different kinds of groups in DNA. A large number of groups are permanently linked to each other in series, forming a long molecule. The order of the groups and the mass-charge configuration of each group provides the basis for complementarity, which is essential for transmitting the information exactly. Thus if the four different groups contained in DNA are identified with the numbers 1, 2, 3, 4, a sequence (or gene) can be generated using, say, 1000 groups in the order 2, 1, 2, 3, 3, 1, 4. . . . Each gene sequence is unique, different from all others even if composed of the same total number of groups. This sequence—the gene—is specific and carries the information for one specific protein. The number of possible sequences, that is, the number of possible genes, is immense and is given by the expression $4 \times 4 \times 4 \times 4 \times$ etc., which is equal to 4^n, where n refers to the total number of groups; 4 refers to the fact that there are only 4 different kinds of groups. Since for a typical gene n is about 1000, it is easy to see intuitively that the possible number of genes staggers the imagination.

In order to ensure an orderly synthesis of the various proteins encoded in a single DNA molecule, each gene sequence is demarcated with "begin" and "end" signals, which indicate the beginning and ending of the sequences that specify each protein. If this were not so, a monstrously large, useless protein would be produced from the entire length of the DNA. The "begin" and "end" signals represent one control element of the system. They are nothing more than fixed locations on the DNA, each with a specific charge configuration. A second element of the control system is a protein that can recognize the "begin" site. This recognition occurs through the princi-

ple of complementarity: the protein must fit the "begin" site
and remain there by virtue of attractive forces. The protein
may be an enzyme (i.e., a molecule that can perform work, say
by joining two molecules)[3] capable of synthesizing a new nucleic acid that is complementary to the gene sequence. This
is the first step in translating the gene into protein. The enzyme
works by joining the four component groups in the proper order. The individual groups have been formed elsewhere in the
cell by another series of chemical reactions, in a manner similar to the one described above. The four separate groups
present in the cellular soup find their partners in the DNA
chain by the principle of complementarity. The groups fit
themselves into position and then are joined together by the
enzyme into a new nucleic acid molecule whose mass-charge
configurations are complementary to one specific gene in the
parent DNA molecule. The new molecule then moves away
from the parent DNA and undergoes a series of interactions
with other cell components, eventually resulting in the synthesis of the specified protein.

But if the cell does not need the protein specified by a particular gene, control elements of the cell can prevent synthesis
from occurring. The cell does this by inserting into the "begin" site a non-enzyme protein that fits more tightly into the
site than the enzyme that would cause synthesis. Synthesis
can start, however, when the blocking protein is removed; this
can be achieved by the appearance of a molecule that can at-
tract the blocking protein more strongly than does the DNA
"begin" site. The attractive molecule is the product of another
series of reactions in the cell, which itself must be controlled.
We can begin to visualize, then, how cellular controls actually
work. There are myriad complex chemical circuits such as I
have described, analogous in some ways to conventional electrical circuits, which are characterized primarily by their interdependence. Each chemical is a control element, controlling a reaction by its presence or absence: the presence of the
blocking protein at the "begin" site prevents nucleic acid synthesis; the presence of the attracting molecule permits synthesis, if the enzyme is also present.

The nucleic acid molecule copied from the gene, as just described, is a "messenger" molecule that acts as an intermediate in a series of reactions that eventually produces a particular protein. Each messenger contains the information from one gene. The messenger itself can do no work; it must first be translated into a protein molecule with the structure originally specified by the gene. The structure (i.e., the mass-charge configuration) of the protein suits it to perform a particular function. Although the messenger is complementary in structure to the gene, and the protein is synthesized from the messenger by a series of steps involving complementary fit, the actual physical structure of the protein is nothing like the messenger or the DNA of the gene. The salient feature is that a segment of DNA comprising a gene gives rise to a specific messenger, which in turn is translated into a specific protein. In molecular biological jargon this transfer of information from DNA to messenger to protein has been called the Central Dogma. An interesting corollary is that the molecular biologist, having cracked the molecular code, can deduce the structure (mass-charge configurations) of the DNA segment in question by observing the structure (mass-charge configuration) of the protein produced from it.[4]

This brief discussion of one of the major interests of molecular biology—the transfer of information and its control—is frankly simplistic, but it represents a good approximation of the state of the art. I might add that about one third of the chemical reactions of a bacterial cell have already been described. Since we know how much DNA is in a cell and how much is required to code for a protein, we can calculate the number of possible proteins per bacterial cell to be between 3000 and 4000.

Although a great deal has been learned about the bacterial cell, the same is not true of the animal cell, which is about 1000 times larger and therefore contains a much larger amount of DNA; the number of proteins is greater and the complexity of the control systems is greatly increased. The study of the DNA of an animal cell can be arduous because it is difficult to separate the DNA containing a desired gene from the large mass of irrelevant DNA; it is a mechanical problem. Recombinant DNA technology offers a convenient and rapid approach to solutions to some of the problems. First and foremost, it can provide large quantities of relatively pure specific DNA segments. This is done by taking advantage of a natural process: the infection of bacterial cells by certain small DNA molecules called plasmids, which multiply there. Recombinant DNA techniques can be used to join a piece of foreign DNA to a plasmid (the "vector") in the test tube. The resulting hybrid plasmid is inserted back into a bacterial cell (the "host"), where it can multiply normally. As a result, the desired DNA segment is amplified many times, along with the plasmid; the bacterial cell is used as a miniature factory, increasing manyfold the actual mass of DNA.

In order to form an unnatural hybrid DNA molecule, or recombinant DNA, DNA molecules from the two different sources are cleaved in the test tube by a special enzyme that creates sticky ends at the site of the cleavage.[5] The two sticky ends at a cleavage point are complementary to each other and to one end of every other DNA segment that has been cut by the same enzyme. Consequently, when DNA fragments of this type from two sources are mixed in a test tube, any ends that are physically close to each other will stick together: they fit each other like a lock and key, having the proper charge configuration to cause attraction.

Some of the joinings will take place between DNA segments from different sources: by this technique, *any* DNA can be joined to *any* DNA. This is a central feature of recombinant DNA technology. The ends that are temporarily joined by complementarity (i.e., stickiness) can then be permanently joined by a second enzyme. Because plasmid DNAs are circular, no DNA is lost by a single cleavage. Both ends are able to join to a foreign DNA segment, thereby regenerating the circle, which is now somewhat larger because of the insertion. The recombinant plasmid is still able to multiply in a bacterial cell. After we allow this to take place, we can retrieve the foreign segments from the new plasmids by treatment with the same cleavage enzyme used originally. Cleavage always occurs at the same sites. The two segments, the original plasmid and the foreign DNA, can usually be separated from each other on the basis of size, yielding a large amount of the foreign segments. If only a single recombinant plasmid has been allowed to multiply, there will be only one type of foreign segment. By repeating this process with many different recombinant plasmids, many pure samples of DNA segments or genes can be obtained, even if the starting DNA is very complex (i.e., containing many types of genes). This procedure is known as the "shotgun" technique.

Although the carrier segment of DNA is usually a bacterial plasmid, the spliced-in segment to be studied may come from any source: bacterial or animal viruses or animal cells. Molecular biologists are most interested in analyzing animal genes isolated by this technology to determine the exact sequence of the four individual chemical groups. These sequences hold the secret of the control elements, and they determine the nature of the proteins produced. It is the aim of molecular biologists to study the structure of all the genes of animal DNA.

Studying the sequence of a gene is accomplished by "looking" at various subsegments of the gene, which are obtainable by treatment with a variety of cleaving enzymes. With appropriate enzymes, the sizes of the subsegments can be made smaller and smaller. Each time, the subsegments can be amplified in amount and purified by the recombinant technique discussed above. In this way the investigator is able to scan the entire sequence of a gene. When the chemical analyses are finished, the results are seen as sequences belonging to each of the subsegments. Finally all subsegments are ordered, as in a jigsaw puzzle, into what is deduced to be the entire gene, including the control elements indicating "begin" and "end" to which I alluded earlier. The sequence is equivalent to a series of numbers such as 34124213. . . .

These examples are not exhaustive; they are meant only to give some idea of why recombinant DNA technology is a valuable asset for the molecular biological study of animal cells. It is easy to see why molecular biologists have generally been ardent advocates of this elegant method of gene analysis. Moreover, they have been so thoroughly seduced by the technology that they have not been able to view the implications of their work in a detached manner. The knowledge gained may

well be useful in eventually understanding the nature of disease, but it would be premature to begin to tabulate the concrete public benefits that might ensue. This is basic research
in largely unknown territory, and its implications can only be
highly speculative. Claims that a cancer cure depends on the
use of recombinant DNA technology in research are a dis-
credit to the scientific community, especially since current in-
dications are that the control of cancer must come at the environmental and life-style level, where most of the causative
factors are found.

Whereas the pursuit of this type of knowledge by recombinant DNA techniques carries a minimum of risk for the general public, provided the research is carried out on a small scale
under strictly controlled conditions, large-scale applications of
the technology are another matter. We cannot blithely ignore
the risk of accident or, perhaps more importantly, the hazards
of success (see later chapters) in applied genetic engineering
without inviting disaster. Many potential applications are already in the gestation stage. For example, in agriculture it has
been suggested that recombinant genes might be able to provide plants with the ability to use atmospheric nitrogen directly; the development of recombinant oil-eating bacteria that
will digest waste hydrocarbons is underway; in the pharmaceutical industry the production of all manner of hormones,
antibiotics and other drugs by recombinant techniques in bacteria is envisaged. At first glance many applications seem attractive and worthwhile. But closer scrutiny is called for to
distinguish real benefits from commercial conveniences that
may have anti-social side effects. I say this at the risk of being
labeled a negativist, for there is much at stake and we lose
nothing by exercising caution. One thing is certain: if recombinant DNA technology follows the path of other technologies,
there are bound to be contraindications.

Risk-benefit analyses of recombinant DNA technology have
been discussed *ad nauseum* since 1973,[6] but unfortunately the
risks considered have been limited to unintended laboratory
events and the participants have been mainly interested scientists. This game of matching accidental risks against benefits
is so speculative that its outcome is strictly a function of the

player. The various congressional hearings dealing with possible legal regulation of recombinant DNA provide a wealth of material not only on substantive issues but also about the nature of the scientific community and its members. It is interesting to compare the disparate views of an economist, several scientists, and a mathematician.

Professor of Economics Roger Noll of the California Institute of Technology has cited a number of basic deficiencies in attempts at risk-benefit analysis.[7] He noted, first, that this kind of analysis is beyond the expertise of molecular biologists. He observed that the scientists had made no realistic evaluation of the commercial feasibility of recombinant DNA technology either with respect to cost or time for development. He said that they also had neglected to consider the risk-benefit ratio for future generations as compared with the present generation. According to Noll, scientists were also guilty of not considering alternate methods to achieve the same ends. Finally, Noll asked: ". . . what benefits from other lines of research by molecular biologists are being sacrificed or delayed by devoting significant resources to recombinant DNA research?"

Professor Bernard Davis of Harvard Medical School offered the following qualitative view, in a report to the House subcommittee on Science, Research and Technology:

"I would like to concentrate on a kind of experiment that is allowed but is causing great concern and is restricted to quite special facilities: the so-called "shot-gun" experiment, in which one transfers random fragments of DNA from mammalian cells. Here it is clear that the probability of isolating a strain with a gene for a toxic product, or with the genes of a tumor virus, is exceedingly low.

"Evolutionary considerations provide an additional and independent approach to the question whether shotgun experiments are likely to create novel and harmful microbes. In my opinion it is exceedingly doubtful that our new-found ability to introduce mammalian DNA into bacteria in the laboratory will create a truly novel class of organisms, for evolution had an earlier crack at the problem."[8]

Dr. Robin Holliday of the National Institute of Medical Research, London, calculated the probabilities of events in a hypothetical scenario of a manmade epidemic resulting from an accident in a recombinant DNA research laboratory.[9] He considered the following probabilities: accidental swallowing of recombinant bacterial by an individual; the recombinant DNA causing cancer in the individual; and the spread of the cancer as an epidemic. He estimated that the probability of one individual dying of cancer from recombinant DNA is one in 100 billion; the probability of a second individual dying is one in 10 trillion; and the probability of a cancer epidemic is one in 100 trillion.

Professor Arthur Schwartz of the Department of Mathematics of the University of Michigan, an expert in probability theory, testified at hearings held by the Senate subcommittee on Science, Technology and Space:

> No responsible advocate of recombinant DNA technology can dismiss the possibility of associated disaster. Instead, eager to pursue research and developments in this fascinating new field, many now argue that if reasonable care is exerted by all engaged in the research as well as the maintenance and assistance crews (a group of workers that may eventually number well into the thousands), if reasonable care is exerted in a system of self-policing, then the probability of calamity is so small that we should not hesitate to proceed.

> As a long-time student, instructor, and researcher in the theory of probability and statistics, I cannot agree with the assertions that the advocates of recombinant DNA research have made about probabilities.[10]

A further reflection on accident probability calculations comes from the field of nuclear power. The instrumentation, hardware, and physical principles in this complex field are well understood, whereas the still more complex variables in recombinant DNA technology are not. Calculation of the probability of a nuclear accident can therefore be made with a high degree of accuracy far beyond the scope of recombinant DNA calculations.

Essential nuclear plant features include an immense assort-
ment of valves, motors, relays, gauges, electrical cables, and so
on. Problems can arise from many sources: design, manufacturing, installation, and construction defects; testing, operational and maintenance errors. Nuclear safety assessments are
further complicated by the subtlety and variety of events that
can arise when one malfunction combines with, leads to, or
induces other malfunctions and creates accident circumstances requiring the emergency operation of one or more of
the plant's elaborate safety systems. There are even more serious problems if there are unforeseen contingencies against
which no protection is provided.

The case of the Oak Ridge Research Reactor accident is
one example of how misleading probability calculations can
be. In this accident there were seven sequential failures, each
involving redundance of three parallel elements, for a total of
twenty-one failures, the absence of any one of which would
have prevented the incident. Three of the seven were personnel failures: an experienced operator threw wrong switches
in three separate rooms; another operator failed to report finding any of these errors; and so forth. The others were design
or installation errors in a reactor with an outstanding performance record. The probability of the event was calculated
to be 10^{-20} (that is, one in 100 billion billion). The event "was
almost unbelievable," but it happened.[11] Again, in the complex nuclear reactor accident that occurred in 1970 at Dresden
II,[12] the most generous assessment of the probabilities of the
separate events could not raise the overall probability above
something like 10^{-18} (one in a billion billion). Yet, here again,
it happened.

The reactor accident at Three Mile Island on March 28,
1979 produced the most serious emergency yet in the nuclear
power industry. The appearance of a hydrogen bubble over
the reactor core was totally unexpected. Harold Denton, director of the Nuclear Regulatory Commission, is quoted as
saying: "[the problem] has not been analyzed. We're into
something that's a different ball game than we expected. The
single thing we may not have anticipated was a buildup of a
gas bubble over the uranium fuel." In its April 13 issue, *Science*

noted that the bubble problem had not been considered in the government's computerized accident simulations. It was pure luck that the bubble did not explode and trigger the "China Syndrome."

Dr. Holliday's calculation that the probability of occurrence of a cancer epidemic is one in 100 trillion (10^{-14}) seems reassuring. He also calculated that "If 10 scientists in each of 100 laboratories carried out 100 experiments, the least serious accident would occur on the average once in a million years." This also seems reasonable in terms of acceptable risks, yet I hasten to add that the probabilities of the nuclear accidents were far, far smaller, and far more accurately determined; nevertheless, they occurred. And that is not reassuring.

At a conference on recombinant DNA risk assessment held in Falmouth, Massachusetts in June, 1977, Dr. Roy Curtiss III estimated at 10^{-16} the probability of transfer of a recombinant DNA from enfeebled research bacteria to normal bacteria (where the DNA would be multiplied indefinitely) within the human intestinal tract.[13] This conference was highly influential in killing federal recombinant DNA safety legislation that had been prepared for introduction in Congress in the fall of 1977; and the following two years, during which recombinant DNA activities have grown enormously, saw no further legislative attempts. But in a letter dated May 11, 1979, sent to the National Institutes of Health's Office of Recombinant DNA Activities, Dr. Curtiss (recognized as the leading authority on these matters) wrote:

... in spite of information presented by E.S. Anderson, H.W. Smith and myself at the [Falmouth conference] and by S. Falkow and colleagues in the published conference proceedings, there was still a degree of concern and uncertainty expressed by the participants on the actual likelihood and consequences of transmission of recombinant DNA from *E. coli* K-12 hosts and vectors to other microorganisms. Since 1977 a number of studies have been conducted which indicate that the overall probability for transmission of recombinant DNA from *E. coli* K-12 hosts and vectors is higher than I or others believed.

... [the new data] would indicate that the cumulative like-
lihoods for transmission of recombinant DNA from EK1
and EK2 [the highest level of biological containment] host-
vector systems are considerably higher than previously believed. Thus, I surmise that if the participants at the
Falmouth conference had been aware of these data, more
consideration would have been given to possible consequences of transmission of recombinant DNA to indigenous
microorganisms of various natural environments. Similarly,
the virologists attending the Ascot Conference [in January
1978] might have also given due consideration to this issue
rather than disregard it in proposing revised containment
categories for cloning of eukaryotic viral information in *E.
coli* K-12 host-vector systems.

Clearly, when it comes to biological complexities, our
knowledge is not extensive enough to yield reliable risk calculations, and those that have been attempted are overly sanguine.

The probability calculations for the risk of human disease
have considered only small-scale accidents in research laboratories. The tremendous exposures that could result from an
industrial accident, for example, where tens of thousands of
gallons of recombinant bacterial cultures could be involved, or
the possibility of design or systems failure on safety facilities
for research, have not been considered, nor the likelihood of
laxness on the part of researchers or technicians. That familiarity breeds contempt for precautions has been documented
by Janet L. Hopson, a science writer who spent ninety-five
days working in Professor Herbert Boyer's laboratory at the
University of California. In an article in the *Smithsonian*
magazine,[14] Hopson reported observing a variety of trans-
gressions of good laboratory practice—all in a day's work with
recombinant DNA.

Dr. Halstead Holman, a professor of medicine at Stanford
Medical School and a primary-care physician, has pointed out
another oversight.[15] He notes that infections of the bloodstream by *E. coli* (the bacteria usually used for multiplying
recombinant DNA) cause a large number of deaths and that

the incidence of infections is on the increase. He believes that three factors contribute to this situation: (1) an increase in the elderly population requiring health care, (2) increased prevalence of chronic diseases, and (3) an increase in the use of drugs that inhibit the immune response.

Holman notes, further, that "much research on recombinants is done in medical centers where there is considerable exchange between the people working in the laboratories and people seeing patients. Sometimes the same person does both. Thus we have a different epidemiological problem from the one envisioned in the guidelines. It is the problem of enfeebled bacteria interacting with persons whose resistance is compromised. It is a question of the epidemiology of infection of weakened human hosts with altered bacteria. Techniques for monitoring and controlling this situation are, at least to my knowledge, not well developed." Nonetheless, facilities for the highest-risk recombinant DNA research have been built in medical centers and highly populated industrial areas. I regard this as irresponsible. The least that could be done is to place these laboratories in isolated areas.

An adequate estimate of acceptable risk is difficult not only because hard data frequently do not exist, especially where fragile or intangible values are concerned, but also because of the fatalistic attitude of industrialized societies that have become inured to dangers and assaults. How often have you heard a cigarette smoker say: "I have to die of something." But one cannot justifiably rationalize harmful acts when other people may have to pay the consequences. It is tiresome and disheartening to hear molecular biologists indulge in this kind of sophistry with regard to the risks of recombinant DNA technology:

> Due to these unknowns [in recombinant DNA] the safest assumption is that adventitious risks and benefits would cancel out, whereas the predictable scientific and practical benefits of gene implantation technology amply warrant use of the technique.[16]

Actually, we can only guess at the nature of the first-order hazards and benefits that may await us. For example, in the early stages of automotive technology, 8,000 cars were registered in the United States at the turn of this century. The worst dangers of the motor car might reasonably have been expected to arise from horses stampeding in panic. A thoughtful response might have been the development of stronger bridle reins and carriage brakes to contain the runaway mishap. Instead, today frightened horses are blameless for the 40,000 to 50,000 automobile deaths a year we suffer in motor vehicle accidents, and for the air pollution of our high traffic density centers. Similarly, at that embryonic stage, predictions on the baneful or beneficial implications of the automobile for private transportation hardly could have appreciated its impact on commercial transport, agriculture, heavy construction, metallurgical industries and warfare. I suggest our attempts to identify the worst and best possible outcomes of DNA technology will appear just as inadequate and be the solution of the wrong problems unless these outcomes are continuously examined and revised.[17]

What I find extremely disquieting is the degree of concern which has been generated in a world where far greater hazards should be everyone's concern. How can scientists accept a world bristling with nuclear weaponry, with the capacity of killing several times over every member of the human race, while at the same time demanding super-safety measures for shotgun or related experiments in heterogenetics? Some biologists would argue that nuclear technology and strategy is not their responsibility; others may even approve of it. To these I would point out that there are other types of biological experiments which are exceedingly dangerous, but which have over the years been carried out in many laboratories throughout the world without generating much concern. I refer to genetic studies with viruses or bacteria already known to be pathogenic to man.[18]

. . . should we forbid international travel simply because our quarantine procedures do not guarantee that exotic diseases will be kept out?[19]

Professor Dworkin of Indiana University, weary of fallacious reasoning, replied to Joshua Lederberg, the Nobel laureate who set forth the last of these arguments:

> That argument, with all due respect, is almost entirely beside the point. If we are remiss about our international travel regulations we should move to correct that situation, rather than taking it as reason for being equally remiss about our approach to the biohazard.[20]

In discussing risks and benefits it is important to identify who bears the risk and who gains the benefit. A manufacturer of useful chemicals who pollutes the river with his by-products gains profit while endangering his neighbors, who may neither use his chemicals nor share his bounty.[21] In recombinant DNA technology, the risks are borne by the community at large, and only they have the right to decide whether the benefits are worth the risks.

4
The Hazards of Success

In its most significant aspect, the discovery of recombinant DNA provides the basis for a new and portentous genetic engineering technology aimed at the creation of unique hybrid organisms according to human design. The use of recombinant DNA could potentially alter man and his environment, for better or worse, intentionally or accidentally. Therein lies both the promise and the danger of this new technology. Unless societal priorities and ethical questions are given searching attention before the events are upon us, the scientific achievement may become a burden rather than a blessing for mankind.

No social mechanism has ever been established to ensure the thoughtful, humanistic application of scientific discoveries. Nevertheless, since 1973 there has been a growing awareness and concern about the implications of recombinant DNA research. Unfortunately, all aspects of the issue save one have generally been dealt with in a cursory fashion; only the biohazardous nature of the research has been taken up in some detail. To be sure, the problem of laboratory safety requires considered attention, and the scientists who participated at the Asilomar Conference, which first addressed this problem, carried out a creditable initial function.[1] But a resistance to discussion of the broader issues has been evident from the beginning.

58

In this chapter I shall look at recombinant DNA not as a research tool but as a technology that will function within a technological society. This is the context in which most of the dangerous aspects of the technique are likely to be manifested. The reason lies in the principles that govern the survival of an economy based on large-scale technology. These principles, which are part of the inner necessity of the system, have little to do with the quality of life or with higher human aspirations. Once a new technology is absorbed into the economy, a large measure of human control over it is lost. This has happened time and again, but our economic dependence on the techno-logical system usually blinds us to this or forces a ration-alization of the facts. Herein lies the most serious danger of recombinant DNA technology. Eventualities that seem too outrageous at the moment even to warrant discussion are liable to become accepted, as necessary evils, after the new technique has become an integral part of the system and thus an economic necessity. This danger cannot be avoided unless we are willing to recognize the fundamental syndrome and anticipate, as best we can, the potential hazards and abuses of recombinant DNA. Only then can we hope to prevent this powerful new discovery from slipping out of our control.

It is unfortunate that discourse on recombinant DNA has focused only on the immediate biohazards that could result from a laboratory accident. Scientists have stuck close to home in their concerns, and few have been willing to risk a conflict between their professional interests and more universal values. When I say that they have been concerned about the biohazards of recombinant DNA research, however, I do not wish to create the impression that they have succeeded in eliminating the immediate risks by the use of physical and biological containment, as contended by some scientists. If a hazard, such a pathogen, exists, physical and biological con-tainment do not "eliminate" it; the hazard is still in the labo-ratory, and the possibility of accidental escape is finite and perhaps greater than one might suspect, as we have seen in the previous chapter. Nor do I wish to imply that maximal pre-cautions have been set up. If they had been, one might expect, for example, that the development of alternate bacterial hosts, ones that do not colonize humans and other vertebrates, would have been a top-priority item; or that the majority of the recombinant DNA experiments funded by the National Institutes of Health would have been directed at the assessment of public hazards. Of the total budget for recombinant DNA research, only a small fraction has been devoted to an assessment of hazards and the development of safer procedures.[2] The implicit assumption has always been that recombinant DNA technology will proceed, using whatever methods are available, regardless of the outcome of any risk-assessment experiments.

In the spring of 1977, legislation for the safety regulation of this research in the United States seemed imminent. A small but powerful segment of the science community was beside itself; something had to be done to prevent congressional encroachment; "freedom of inquiry" had to be upheld. Never before had so many influential members of the science community felt the call to political duty so strongly; out came the big guns. This revered community was now to behave like any other special-interest group. Presidents of learned societies began to knock on doors in Washington. Frantic attempts were made to dress up fragmentary and preliminary data on risk assessment so that recombinant DNA technology would seem to entail negligible risk for laboratory workers and the public. The aim of these efforts was clear: to undercut pending federal legislation and eliminate public participation in decisions regarding the safety of recombinant DNA activities or other scientific matters. (I discuss this fully in chapter 7.) The science community won its battle: the legislation was watered down and finally killed. Congress and many members of the scientific community had been diverted from the real issues by those who erroneously see laboratory safety legislation as a threat to freedom of inquiry.

In the long run this monumental lobby will have the unfortunate effect of discrediting the scientific community. If the original discourse had been conducted on a broader basis, taking into account the moral, ethical, and technological aspects of the issue, the discussants would have realized that the primary problem is not scientific but technological. We would not find ourselves today polarized into the scientific opponents and proponents of recombinant DNA. We would see quite clearly that scientific freedom is not really what is in question, and that regulation belongs primarily at the technological and not the scientific level; that scientists have played into the hands of industry by stifling attempts at legislation rather than helping to create meaningful regulation where it is needed most. They have confused freedom of inquiry with freedom of technology.

As a consequence, more than seven years after the first warnings about the hazards of recombinant DNA, the public is offered very little assurance regarding dangers arising in the laboratory. What remains of the safety guidelines developed by the National Institutes of Health are enforceable only by the punitive action of withholding NIH funds for research. Of course, no such restriction is possible with private financial sources, such as those of industry. Furthermore, as pointed out by the Senate Subcommittee on Science, Space and Technology in a 1978 report on its oversight hearings,[3] the NIH guidelines do not fully insure the accountability of research institutions and investigators, and the standards do not purport to deal with prospective commercial applications. The Subcommittee made specific recommendations for rectifying these and other deficiencies. Whether these recommendations will ever be realized by future legislation remains to be seen; right now it looks doubtful. No legal mechanism has been established whereby the public can take part in a meaningful way in the debate, and there has been only token mobilization of input on the recombinant DNA issue from those who are particularly qualified to consider its social and ethical aspects.[4] Industry is free to do anything it wishes, although the subcommittee suggested using existing statutory authorities for premanufacturing and premarketing reviews. Thus, in spite of the publicity surrounding the National Institutes of Health safety guidelines, there is little but voluntary precaution to protect the public from the danger of laboratory or industrial accidents. The controversy in this area has tended to obscure a much more serious omission: the lack of any mechanism for long-term assessment and control of commercial applications. This subject has barely been touched upon in the recombinant DNA safety debate.

Meanwhile, research aimed at industrial applications has proceeded with a speed rarely achieved in molecular biology.[5] Many new companies have been set up to exploit the technique, which is likely to become industrially entrenched in record time. A number of molecular biologists doing academic recombinant DNA research have obtained patents and have also become members of corporations interested in applying the technology. Eli Lilly and Company has entered into a multimillion-dollar agreement with Genentech, a genetic engineering firm, one of whose main officers (Herbert Boyer) was the first scientist to produce a mammalian product (the hormone somatostatin) in bacteria, using recombinant DNA techniques. Genentech has also synthesized human growth hormone, as have scientists at the University of California at San Francisco.[6]

The announcement that another new company, Biogen, has developed a recombinant microorganism that synthesizes small amounts of a substance closely resembling human interferon, a promising but still speculative mammalian antiviral agent, caused an immediate jolt in the stock market and a burst of glowing publicity in the public press—although it will probably be some years before the material will be ready for clinical testing to determine its usefulness. The Upjohn Company is trying to establish the legality of patenting recombinant bacterial strains for commercial use.[7] The potential commercial pay-offs appear to be so close to reality that several other major corporations are also investing heavily in the development of recombinant DNA techniques. A transnational company has been set up by a Canadian multinational firm, employing U.S. and European scientists to carry out recombinant DNA research wherever it is most convenient. A major stumbling block to commercialization of the new process in the U.S. is the physical size of the production volumes. The NIH guidelines call for a voluntary ten-liter (about 2½ gallons) limit, far too small for commercial purposes, which will require thousands of gallons. However, Eli Lilly and Genentech have asked for and received permission to scale up their procedures for insulin production.

Drugs are not the only products seen as possibilities for recombinant DNA research. A scientific-industrial revolution is on the way. Conversion of biomass to alcohol, production of ethylene for the plastics industry, the use of bacteria to facilitate the extraction of metal from ore, new agricultural technologies—all are being seriously considered by various entrepreneurs. Some are already under development. No investigation of the possible side effects of these technologies, some of which could alter vital ecosystems, has been reported. If we do not act now it will soon be too late, if it is not already too late, to direct the way in which this uniquely powerful scientific achievement is to be used.

It is not realistic to expect industry, however responsible it may be, to show more concern than the public or its elected representatives about the long-range implications of the new technology. There are plenty of recent examples to illustrate the results of this sort of technological laissez-faire: the development and use of various pesticides, food additives, asbestos insulation, polluting fuels, and a host of other agents clearly demonstrates that success in solving an immediate problem is sufficient for the establishment of a new technology. So far, it has not been necessary to show that the new technology will not create problems worse than the old ones. Let us consider one imminent application of genetic engineering. There is a strong impetus to design a bacterium capable of consuming oil inadvertently spilled by faulty oil tankers on the oceans of the world; on the bacterium under way at General Electric. When an appropriate organism has been developed and high oil interests are clamoring for it, who will decide whether it is safe to pour carloads of these bacteria into the oceans? Is there sufficient knowledge to be able to predict all the consequences? Will the oil companies or General Electric be strongly motivated to preserve the ecology of the oceans, which belong to all of us?

When released, the oil-eating bacteria will no doubt perform their task as designed, with great success. Any incentive to take precautions against oil spills will decline. Meanwhile, the release of vast quantities of one organism, and its petroleum and other breakdown products, will constitute an assault on ocean ecology. One need not know details about specific chemicals; the sheer mass of material, repeatedly applied, will be enough to disturb the equilibrium of aquatic life. The oil pollution problem will not be eliminated; it will simply be transmuted into another kind of pollution, the consequences of which cannot be fully tested in advance because we do not know enough about the complex interrelationships of life in the ocean to set up an adequate test system. But the unique aspect of the problem is this: if the newly-designed bacteria should find an unforeseen ecological niche, there could be long-range and almost certainly irreversible consequences, which might not become evident immediately. Thus the success of the oil-eating enterprise is inseparable from a number of monumental risks. In fact this is a fundamental characteristic of many modern technologies: their very success spawns new problems—the hazards of success. While this and other revolutionary new projects are gestating, we should be preparing a mechanism for independent review and assessment of proposed applications of recombinant DNA technology, particularly with respect to their future impact on human beings and their environment.

Let us stop and think about the question of reversibility for a moment. Recombinant DNA technology produces organisms with new gene combinations. If they find themselves in an appropriate environment, these organisms will replicate and become permanent residents in our ecosystem. There has been much reassuring talk about the fact that recombinant bacteria usually show decreased viability. This is true in most cases, but it is not inevitable. The aim of genetic engineering is to design organisms that will thrive under a particular set of conditions—on an oil slick, for example, or in a manmade environment. In order to be useful, they must be viable under these conditions. Some of these recombinant strains are likely to find suitable growth conditions elsewhere as well; there is incredible ecological variety in nature. Even if it is normally confined, any strain used on a large scale will certainly find a chance to escape and look around for a home in the wild. (Remember the nuclear spills, and how we were reassured about the safety of nuclear power plants?[8]) Some strains, such as the oil-eating bacteria, will be released intentionally. With the number of laboratories working with recombinant DNA increasing exponentially throughout the world[9] and the countless types of experiments, and with the imminent expectation of large-scale industrial applications, it would be naive and perhaps disingenuous to create the illusion that no organisms capable of competing successfully in nature will be produced. Once these organisms find their niche, they will be with us for a long time indeed; they will be irreversible.

Recombinant DNA technology will couple this organic irreversibility with the irreversibility of modern technology. Although we naturally assume that undesirable industrial processes could be terminated if we so chose, this is true only in part. In the context of our present socioeconomic structure we could not terminate the manufacture of automobiles even if we wanted to; automobiles are an integral part of our social structure, and their absence would portend a virtual collapse of our economy. John Lear has neatly questioned the phenomenon of the automobile:

It is argued that anxieties about the future should be avoided. That argument would leave the future to chance. The future was left to chance at the time Henry Ford made the automobile a household commonplace. What have we now as a result? Fertile farmland paved over, closed to the planting of food crops. Rainfall runoff from the pavement periodically flooding sewage treatment systems and polluting streams we depend on for drinking water. The purity of the air we breathe polluted by automotive exhaust fumes. Sunlight acting on the fumes, generating smog and altering the climate around big cities. All of this ugliness might not have been foreseeable in Ford's day, but the logic of it is so straightforward that much of it surely would have been at least suspected, if possible sequelae of the motorcar's advent had been considered then along with the pleasures of personalized transportation.[10]

Again, in principle we could stop the manufacture of

cigarettes now that they are known to be harmful, but in practice this seems to be impossible. The economy would not be seriously threatened, but expert advertising has created a demand for cigarettes in spite of the incontrovertible causal relation of tobacco to cancer and heart disease. Cigarettes are here to stay; they are "irreversible," and their use is increasing. This points up the fact that social and economic forces are sometimes more powerful than scientific facts, and emphasizes that even a nonessential and patently inurious activity cannot easily be eradicated once it is securely ensconced in the system. In our technological society, this is a fact of life that must be accepted. As President Carter said in a recent address to tobacco farmers, he sees "no incompatibility" in his administration's policy of supporting tobacco prices and at the same time pressing a vigorous campaign to warn Americans about the dangers of smoking. On the same day, the *New York Times* reported on a study by the American Medical Association which concluded that cigarette smoking can cause irreversible heart damage and might be responsible for maladies ranging from indigestion to cancer.[11]

Recombinant DNA technology will impose a double jeopardy: the irreversibility of the organisms themselves and the irreversible socioeconomic entrenchment that will result from the successful use of recombinant organisms, regardless of their side effects. The irrationality of the system is such that even if it could be proved in advance that the use of oil-eating or drug-producing bacteria would have catastrophic consequences, this would very likely not prevent them from becoming a commercial reality—as long as the disaster was not expected to be instantaneous and massive. In a technocracy everybody is resigned to the fact that if something *can* be done, it *will* be done. A sensible solution to this dilemma would require a set of values and priorities based on fundamentally different economic concepts, ones that do not require continued growth and expansion.

One reason for the inexorable drive to incorporate new techniques into the system is the need to provide technological fixes for past failures that cannot be rooted out at the source. Thus we try to find a technique for curing lung cancer while we continue to manufacture and advertise cigarettes, and we develop oil-eating bacteria to clean up oil spills instead of re-designing oil tankers or reexamining our energy-intensive and wasteful economy or making a serious effort to shift to re-newable and ubiquitous energy sources. Many of the benefits expected from recombinant DNA technology are similar to this. Technological fixes have become such a familiar class of activities, such an integral part of everyday life, that they are hard to distinguish from solutions to problems arising from *real* human needs. The cancer problem is a stark case in point. The $1.2 billion spent on cancer research in 1977 represents in large part a search for some means to patch up the damage caused by environmental factors, including industrial carcinogens and agents such as food additives.[12] Members of Congress and the National Institutes of Health feel justified in this approach; they think they are giving the taxpayer his due. The real solution—to eliminate or reduce environmental fac-tors that cause cancer—is largely neglected. A leading cancer expert, Sir Richard Doll, has said that "most if not all cancers have environmental causes and can in principle be prevented."[13] But it seems to be taboo even to *think* about such a rational approach, because it implies an attack on our way of life. Because of the insidious assumption that environmen-tally caused cancer is an immutable fact of life, the search for a cancer cure is not recognized by most people as a technolog-ical fix but as a humanistic activity.

The now familiar list of potential benefits that may accrue from recombinant DNA includes, according to the Environ-mental Impact Statement of the National Institutes of Health: the production of insulin, the production of antibiotics, the production of vitamins and hormones, the production of oil-eating bacteria, and the increased production of food crops. A cursory analysis of these potential benefits prompts the ques-tion: Do we need them? Let us examine the case of insulin. At a meeting of the National Academy of Sciences held in March 1977, a spokesman for Eli Lilly stated that there is no insulin shortage at present.[14] When I spoke to Dr. Herbert Boyer at the California hearings on recombinant DNA safety in Janu-ary 1976, he said that the cost of insulin produced by recombinant DNA techniques developed by him would be about the same as that of insulin obtained from cattle. Furthermore, bovine or porcine insulin is satisfactory for human use, and reactions to the material as a foreign protein do not present a serious problem.

A thoughtful approach to the problem of diabetes and the use of insulin for its treatment was given by Harvard's Professor Ruth Hubbard at the 1977 NAS meeting.

What I am suggesting is that what we need to know in order to study the cure for diabetes are the causes of diabetes, which are, as with all other diseases, heavily influenced by social and environmental factors. This is not to downgrade diabetes as a health problem. It obviously is; it is among the top eight killers in this country. But we need to know more about its real causes, and the real causes are not lack of insulin. So the thought I want to leave you with is this: before we jump at technological gimmicks to cure complicated diseases, we first have to know what causes the diseases, we have to know how the therapy that we are being told is needed works, we have to know what fraction of people really need it. There are lots of questions that we have to answer in order to lick diabetes, and diabetes is a major health problem. But what we don't need right now is a new, potentially hazardous technology for producing insulin that will profit only the people who are producing it. And given the history of drug therapy in relation to other diseases, we know that if we produce more insulin, more insulin will be used, whether diabetics need it or not.[15]

The need for making insulin by a new method is therefore unclear on scientific and humanitarian grounds. In all probability the producers are attracted to recombinant DNA technology because it may be expedient in the future, and insulin production will make a convenient pilot project. The situation is essentially the same with respect to the use of recombinant DNA techniques to produce antibiotics, vitamins, and many other proposed products. It is hard to find any justification in these examples for subjecting the public to even minuscule risks.

The proliferation of efforts unrelated to human need, in the drug industry as elsewhere, results in the dissipation of resources needed for more important purposes. A recent report released by the World Health Organization has indicated that only 210 drugs would be sufficient to fill world health needs. Yet in affluent countries such as the United States the report noted 20,000 pharmaceutical products in medical use, corresponding to 2000 to 3000 substances.[16]

Drugs are only a small fraction of the chemicals currently manufactured; over 4 million chemicals are produced, of which more than 3 million are organic chemicals. The U.S. Environmental Protection Agency has the job of keeping track of them. The EPA estimates that about 50,000 chemicals are in everyday use, and this does not count pesticides, pharmaceuticals, and food additives. The Food and Drug Administration estimates that there are about 4000 active ingredients in drugs (WHO estimates 2000–3000) and 2000 that are used as stabilizers. FDA also estimates that 2500 additives are used for nutritional value and flavoring and 3000 are added to promote product life.[17] The EPA estimates that there may be as many as 1500 active ingredients in pesticides. The mass of work involved in testing all these products has prevented the work from being accomplished in a reasonable time, with the net result that the public is at present unprotected.[18]

If the public is to be subjected to any degree of risk, surely the risk should be justified by real public benefit. The use of recombinant DNA technology for agricultural purposes that allegedly would help solve the world food problem sounds like one of the more worthy applications. But we must not let our understandable sympathy for the hungry people of the world lead us into mistaking the cause of the problem, which is not one of production or quality but of distribution and utilization. The world now produces enough grain to feed everyone adequately; but in the affluent countries, grain is actually wasted in amounts that *far* outweigh the needs of the hungry world.[19] In addition, many countries with undernourished populations devote their agricultural resources to the production of coffee, bananas, or other nonessential crops for export. These are economic problems, not scientific ones; they require a political solution, *not* recombinant DNA.

No real need has yet been brought forward to justify the serious ecological hazards of introducing major disturbances into the complex balance of living things. Although it is true that we have been disturbing natural systems ever since we invented agriculture and domesticated animals, it is a question of scale, and it is a question of irreversibility.

I have pointed out that after a product or device has become an integral part of the socioeconomic structure, it is usually too late to alter it in any meaningful way even when the risks and hazards far outweight the benefits. I have tried to show that, in addition to the palpable risks of accident, the hazards of successful recombinant DNA technology are very serious indeed, whereas many of the benefits proposed do not correspond to real human needs. I suggest that what is most needed now is an all-out effort to identify the true needs that could best be filled by recombinant DNA techniques and to study the risks they entail. Only when the risks are adequately understood can they be weighed against the risk of not taking action, or of taking alternative action. If, on the basis of a full analysis of risks and benefits, certain benefits are judged by a disinterested and representative body to be worth pursuing for the common good, then let us concentrate on these particular uses of the technology and eliminate the others. And let us proceed under strict, legal safety controls, with continuing study of the future implications of what is being done.

5
Science as Technology, and Vice Versa

The human hormone somatostatin was the first functional protein to be synthesized by gene-splicing, and with its synthesis for the first time in the field of molecular biology, science and technology coalesced. The scientific goal was to show that a human gene could be expressed (i.e., make protein) within a bacterium, whereas the technological aim was to show that it is feasible to manufacture a protein in this new way for the commercial markets of the world.

The feat was announced to the public by the president of the National Academy of Sciences when he testified before the Senate Subcommittee on Science, Technology and Space in 1977.[1] The Academy's Statement breached one of the canons of scientific propriety; the work had not yet appeared in a scientific journal, where research results are normally subjected to the scrutiny of peers. Release of unpublished information through inappropriate channels has been frowned upon by the scientific community, at least until the recent past. A top-ranking journalist from the old school was so disturbed by the Academy announcement that he brought it to the attention of scientists in a letter to *Science* magazine;[2] he called for discussion, but none was forthcoming because *Science* refused to publish letters on the subject, perhaps for fear of bringing this breach even more into the open.[3]

Evidently the president of the National Academy felt that

the synthesis of somatostatin presented an elegant opportunity to squelch any new antiscience feeling among a citizenry long since overwhelmed with technological items, some useful, some not. Here was a human hormone holding great promise, coming very soon after the birth of recombinant DNA technology. The announcement was propitious because it was expected to modulate the intensity of the contemplated federal safety legislation. In effect, the National Academy of Sciences implicitly condoned a break in the tradition of scientific publication in order to gain an advantage in public relations.

But more importantly, the National Academy announcement signaled a changing relationship between science and society. The scientists who synthesized somatostatin, themselves acted out the familiar technological dictum "if it can be done, it will be done," without allowing for any reflection about its ultimate public value. They took it into their own hands to proceed with a commercially useful application of this new and powerful dangerous technology before the creation of institutions to control its applications in the public interest. Science, in this instance, was practiced as technology, and in the end, the practice of science will suffer.

Somatostatin was just the beginning of the story. At about the time the Academy's announcement was made, at least four laboratories were working concurrently on the bacterial synthesis of insulin, another mammalian protein hormone. Clearly, keen competition had developed among the several laboratories. A news release appearing in *Chemical and Engineering News*[4] at the time noted that "a team of scientists led by Dr. Walter Gilbert of Harvard University has just *edged into the lead*" (italics mine). The language of this article treats science as a bona fide tournament. This is regrettable because it tends to legitimatize this sort of fierce rivalry, fostering secrecy and noncommunication—hardly a desirable feature in science, where the exchange of ideas should be encouraged, for science is a cooperative edifice whose development depends exclusively on adding bits of knowledge to the existing structure. In the race for insulin, science was converted into something resembling a commercial venture. And what for? There is no crying need for insulin;[5] evidently the scientists had their eyes on worldly goals.

The academic scientists who first synthesized interferon also had their sights set for practical returns rather than pure science. Their results were released at a press conference, not in a scientific journal that would have demanded experimental details. Nicholas Wade of *Science* magazine commented on this maneuver by noting that the stocks of Schering-Plough and Inco, which own shares in a company with which the scientists are associated, rose the day after the announcement. This and similar nonscientific announcements of scientific results have led even some strong proponents of recombinant DNA technology to complain bitterly of the industrial connection.

This is hardly the quiet, thoughtful research activity ordinarily associated with the pursuit of pure knowledge. The practice of horseracing tactics in high places induces other scientists to follow suit. Why not, if Harvard can do it? Moreover, the more they read about such tactics in the press, the more the unsuspecting public will come to accept this behavior as the norm, for they have no basis for rejecting science as technology. Indeed, the public, by comparing science with free enterprise economics, may suppose that "competition" is desirable. In a society in which science and technology have come to be so closely coupled, it is easy to overlook the fact that science is a search for truth, which cannot be made "better," and that its results are not always commercially useful.

The admiration of our technological society for the efficient and single-minded pursuit of a clear-cut goal inclines it to try to jam science into a technological mold. The National Cancer Act of 1971, mentioned earlier, was conceived of as a crash program, similar in principle to a moon landing. But basic scientific knowledge was in no way adequate for a technological approach to the cure of cancer. For a number of years, funds were lavished on mission-oriented programs that have not (and could not have) lived up to the expectations of Congress and the public.

Science practiced as technology is one of the symptoms of Big Science. How did Big Science come into existence? The transition was brought about by four major factors operating simultaneously. First, in order to maintain the continuing expansion of research efforts started during World War II, high-level government officials were convinced first by physicists, who reminded them of the success of radar, and then by biomedical scientists, who talked of penicillin, that wonderful things could come from scientific research. Funds for research became available in amounts that were at first astonishing, but soon became the accepted order of the day. Second, scientific results converged from different directions in the postwar period, and several new and basic discoveries were made that served as powerful catalysts to a still deeper inquiry into biological structure and function. Third, again as a result of wartime research, new and sophisticated electronic instrumentation came into the scientific laboratory, representing one of those occasions in the science-technology interaction when technology aided science, rather than vice versa. Fourth, after the austerity created by World War II, the public psychology was geared to seek the good life, using the means that had been so successful in the war effort.

In health-related fields, the arrival of the sulfa drugs, followed by penicillin, ushered in a magical biomedical era. Overnight the use of penicillin led to the essential eradication of bacterial illnesses; the wonder of this achievement created a deep patronage toward the scientific community. When it was learned that there were allergic reactions in some cases and that resistance developed in others, the pharmaceutical firms were only too willing to develop new derivatives or formulations. In the meantime, innumerable new antibiotics were introduced on the market; the more the better. As frequently happens, technology itself became scientific, developing chemical and biological approaches for the specific purpose of improving and inventing pharmaceuticals. When the setting is a commercial establishment, one tends to label the work as development, whereas the same work in a university would be termed scientific research. This only underscores the intimate relation of science to technology: the social setting often determines the terminology.

Basic research was also carried on in the area of industrial chemicals. Dr. Caruthers of the E. I. Du Pont de Nemours research laboratories was doing pure research in the early 1930s when he examined the properties of what are known as high polymers. These are organic chemicals characterized by their very large size, which enables them to coalesce into solid matrices of any desired degree of plasticity: plastics. His work, which can easily be classified either as science or technology (though I am sure he did not think so), resulted in the commercial production of the first important synthetic plastic, nylon. Du Pont reminded us of its industrial origin when it told us that nylon was made of coal, air, and water. (Remember the motto? "Better things for better living, through chemistry.") How better to instill even more awe of science? The result: more and more plastics, as well as a vast array of other chemicals—detergents, pesticides, food additives, and many more.

But what was to be the effect of this renaissance of technology-related science? Congressmen with their ears to the ground felt the tremors. Having been introduced to science and technology through the Manhattan Project, senators and representatives were on the proper wavelength. They did not have to wait for some of their number to die of cancer or heart attacks in order to implement basic research programs in health-related areas. The National Institutes of Health, which started as a small operation in 1930, soon became the recipient of comparatively large sums of money. By 1977 their combined budgets had risen to about $2.5 billion. Other agencies, such as the National Science Foundation (created with miniscule funds in 1950, it had a budget of $750 million in 1977) and the Atomic Energy Commission also instituted programs in basic research in health-related areas. By the mid-1950s we were well on our way to Big Science; the Sputnik affair in 1958 injected new vigor into the effort, which soon put to rest any lingering doubts that science had become Big in America.[6]

With the influx of federal money into basic research, the process of mission orientation began. The federal government had long been in the practice of using the contract system to accomplish its technological aims, and it was known even in the immediate post–World War II period that government-sponsored research was expected to be mission-oriented (sometimes called "targeted research"). However, the format of grant applications was broad enough to encompass a spectrum of scientific activities, and no one felt that he was "cheating" if the original plan was not strictly adhered to; for how could a scientist not follow a "hot" lead, even if it was not related to the original proposal? Thus it became customary for scientists to pay little heed to the research goals they were obliged to set forth in their grant applications.

In effect, there was substantial "freedom" of choice, and the choice did not then have to be made surreptitiously. But, with time, the federal reins began to tighten: the aims of research had to be chosen from a narrower field and had to be adhered to much more closely, resulting in less freedom for the individual investigator.

This loss of scientific freedom was so insidiously slow that at first it was hardly perceptible. But in the late 1960s federal cutbacks signaled starkly that the golden era for biomedical research had ended.[7] The funding situation was exacerbated by exorbitant increases in the cost of equipment, increases that were several percentage points higher than the general rate of inflation. Scientists now had not only to scramble for subsistence funds but to buckle under and produce more and more. A bitter pill indeed.[8] These days, many a scientist is only too happy to become a public servant by performing targeted research.

But, in fact, has he really become a public servant? For the most part, no. The projects and programs that have been foisted on scientists were developed by scientists and science administrators responding to actions or directives that originated in high places in government (and were often based on wishful thinking). For example, the Special Virus Cancer Program was set up in 1964 (1) to determine whether viruses are agents involved in human cancer and, if so, (2) to develop means for prevention and/or control of human cancers. The 1974 Zinder report (by an *ad hoc* group of scientists responding to a request of the National Cancer Board), after noting that a quarter of a billion dollars had been spent from 1964 to

1974, criticized the basic premises of this program, stating that "there did not, nor does there exist, sufficient knowledge to mount such a narrowly targeted program. The same two objectives [noted above] remain."[9]

One reason for the perpetration of mistakes of this kind has been pointed out by the Interdisciplinary Cluster on Biochemistry, Molecular Genetics and Cell Biology of the President's Biomedical Research Panel, which states in a recent report:

> Major decisions on fund allocations for biomedical research are generally made without adequate or sufficient advice from the scientific community; this applies to every step in the formulation and execution of the budget. . . . The initiative belongs to the Office of Management and Budget. . . . In effect all important decisions concerning fund allocations for biomedical research (among Institutes and in each Institute) are made by a small group of budget makers from OMB and administrators or scientific administrators from NIH staff.[10]

The scientist is thus not responding to the needs of the public directly, but only as those needs are seen through the eyes of a few managers working in a technocracy. Those managers may have some societal needs in mind, but it is also clear that the entire system of assigning priorities for scientific programs and allocating funds is highly influenced by expediency and by pressures from lobbyists.[11] This kind of science hardly compensates for the scientists' loss of freedom.

It must be plain by now that in a world of science practiced as technology the vision of a lone scientist in a white coat, working tirelessly and persistently to achieve his purpose, is not part of the present scene. Modern science is schizoid: mission-oriented aims *versus* intrinsic interest; the hungry giant that pays for science *versus* the scientist's inner hunger. This dichotomy causes a conflict for the scientist, who feels himself torn between the ivory tower of pure research and the real world that expects tangible results. But the ivory tower is more illusory than ever. The scientist rarely has the opportunity to pursue the research of his choice in whatever direction it may lead; that privilege is out of date. If he is lucky he can manage to link some of his own interests with the project at hand; he re-tailors his science periodically so that it will have the appearance of being useful to the taxpayer and at the same time will satisfy his thirst for pure knowledge. In an unusually frank article in the British science magazine *Nature*,[12] a scientist recently observed that the investigator who seeks scientific information for its own sake rather than for a specific applied purpose has to do so in "bootleg" fashion; that is, he must quietly divert some of his funds and time from mission-oriented experiments projected in his grant application to experiments that, in his judgment, will enrich science more.

The administrative tasks of procuring research funds and writing progress reports consume more and more of the investigator's time. In 1978, 47,000 proposals were submitted to the principal U.S. government agencies; this represented 2700 man-years for the writing and at least 3300 man-years for the review.[13] There is no letting up; the restrictions placed on the scientist and the demands he places on himself intensify every year. He also realizes that the practice and the management of science, entirely different operations, are inseparable and that both fall on his shoulders. Moreover, the contemporary scientist is more than ever aware that in order to stay in the business of science he must produce; and it helps immensely if his discoveries are laced with public relations. The scientist nowadays feels, almost every morning when he goes to work, that this kind of science is not what he had in mind when he became a scientist. Especially for one trained in the days of Little Science, the feeling is so intense that it borders on frustration. The problem is that the scientist has been misled. He has been taught to expect that in his work he will be able to pursue truth, but in fact he may find himself carrying out a mission whose aims and rationale he cannot truthfully appreciate. He may even dislike the project. Furthermore, next year he may be on a different mission, the end point of which may be equally questionable in his eyes. The scientist would be less frustrated and perhaps experience some degree of fulfillment if he were permitted to participate in the analysis of societal problems and to become part of the decision-making process. But this seems to be a vain hope in the present milieu, where science is often practiced as technology and decisions frequently do not arise from considered judgments.

The growth of federal funding after World War II not only resulted in an enormous increase in trained, highly paid scientific personnel, but also in the acquisition of new and sophisticated research instrumentation. As a result of major discoveries like the transistor and the design of new electronic equipment arising from wartime research, a new genre of instrumentation was available to the scientist. He could carry out his experiments more efficiently and more accurately in a shorter period of time. What is more important, he could perform measurements not heretofore possible, thereby opening new domains of research. Among the first of these touchstones were the radioactive isotopes that were a direct result of the Manhattan Project. These chemical elements are chemically identical to the everyday variety except that they emit radiation and can be detected easily by appropriate devices. Ordinary iodine found in table salt is an example. When it is radioactive, iodine can be traced throughout the body very easily. The amounts required are vanishingly small, and this is the central feature of the use of radioactive elements. They can be "seen" and traced, making it possible to do a wide variety of measurements that ordinarily would require such large masses of material that the system under observation would be perturbed out of normalcy.

Another quantum jump forward in biological research was the electron microscope, a product of physics and engineering research. Using wavelengths of electrons instead of ordinary light (as in the common optical microscope), the electron microscope permitted visualization of atoms and molecules for the first time. Actual photographs of proteins and nucleic acids were easily made, revealing details of structure not previously accessible. In another area, laser beams, which are a source of intense coherent light, made possible a new type of light-scattering experiment that enables the scientist to probe more deeply into the structure of molecules, such as those in plastics; laser beams can also be used in microsurgery and metallurgy, where intense sources of energy in very small areas are required. In the realm of applied mathematics, high-speed computers permit calculations in nanoseconds and less that might formerly have taken inordinate lengths of time, measured in years; space technology as we know it today would not exist without the computer.

Enticed, and understandably so, by the availability of such sophisticated electronic equipment, scientists have become accustomed to a new kind of experimentation, one in which results are increasingly seen on dials and printouts rather than as concrete substances. This experimental refinement costs the scientist more than money; he pays with his dependence on instruments that can be repaired and maintained only by expert technicians at their convenience and at exorbitant costs. This represents a camouflaged superteam effort: a team (or teams) in the research laboratory coupled to a commercial team (or teams) supplying the instrument. Moreover, instrumentation itself has become scientific, with the result that improvements embodied in new models are almost yearly events. Scientists feel at a disadvantage if they cannot avail themselves of the latest technology in instruments. But instrumentation tends to reduce many scientists to high-level technicians. Team effort has the same consequence for the scientist as mass production has for an assembly-line operator: the separation of effort from end product, with its psychological price. In addition, the necessity or urge to produce scientific results in a hurry leads in many instances to an almost frenzied atmosphere. In this milieu the scientist often passes up the opportunity to follow a new lead, frequently manifested by an inconsistent result, a fluke. This is a profound loss, for that is the way much "pure" knowledge is born. The mass production of contemporary science is not really efficient.

Another major factor that helped impel biomedical science forward came from a number of major discoveries that profoundly altered scientifc thought and theory. These came to light during a period of about ten years, 1945–55. The discovery by Watson and Crick of the structure of DNA, the now well-known double helix, laid the foundation for molecular biology. Prior to this, the uniqueness of the physical structure of DNA had not been appreciated, in spite of much effort directed to the study of DNA in the previous decade. The Watson-Crick hypothesis was the start of a new era in biology. The discovery of bacterial viruses as unique entities composed of DNA (or RNA) and proteins also introduced biological science to a new domain: the study of viral infection and its molecular consequences. This work was a cornerstone in the foundation of molecular genetics because viruses, being small and relatively simple, could be far more easily analyzed than cells with respect to their molecular and genetic characteristics. X-rays were shown to alter the genetic substance (DNA) of living cells. These alterations, known as mutations, could lead to a different behavior of the cell—for example, causing it to become cancerous. More important than the specific effect was the fact that, through DNA, the heritable characteristics of the cell could be altered in a permanent way. This work led to a search for the means of producing specific and heritable changes in living systems.

The collective set of new discoveries made possible for the first time in the history of biology an understanding of many biological functions at the molecular level. More important, the framework for understanding a great deal more about biology now came into clear view. Up to this period, morphological descriptions (i.e., descriptions of the sizes and shapes of cellular components), as revealed by the ordinary light microscope, constituted a major part of experimental biology. Now it was possible to go further. To understand the inner workings of a living cell at the molecular level necessitates a molecular dissection of the cellular components, which are chemical substances; intricate chemical interactions cannot be studied in an intact cell. Therefore the cell must be cracked open, its viscera extracted, and its components subjected to detailed examination; such an investigation represents an ultimate analysis. Herein lie the seeds of reductionism.

Reductionism commands a high price: the inevitable loss, implicit or explicit, of the meaning of wholeness. This way of thinking, the theoretical isolation and study of inseparable attributes, is one of the intangible legacies of science to the technological society. Theodore Roszak[14] and other social critics have pointed out how deeply reductionism has permeated modern thought and influenced our way of life, causing much of the fragmentation and alienation found in all sectors of industrialized societies.

The resurgence of science after World War II has had pervasive psychological as well as practical consequences. Industrialized society has become virtually addicted to the practical applications of scientific research. The affliction is due in large part to the pushers of a wide spectrum of wares; whether we need new medicinals or not, we are plied with a plethora of natural and synthetic drugs. Physicians are deluged with commercial brochures describing the latest innovations, which for the most part are variations in formulations of drugs already in existence. It takes an expenditure of between $7 million and $10 million and years of laboratory and clinical work, to develop a new drug—an inefficient use of scientific resources, at best. As I pointed out earlier, the World Health Organization has indicated that at present there are about one hundred times more pharmaceuticals on the market than are needed for adequate medical care.[15] The development and marketing of new formulations also cost money, but it is in the manufacturers' interest to create the illusion that "new" is necessarily "better." In this charade, technology masquerades as biomedical science.

To separate facts from advertising fiction would require careful analysis and scrutiny, for which no one has the time or the sense of urgency. As a net result, the public is duped. And so thoroughly duped that it is virtually impossible to convince it otherwise. In a way this is understandable. Having introduced the Pill as a convenient and virtually foolproof method of contraception, for example, it will take more than logical arguments to dissuade those who have come to rely on it, and to establish a more rational approach to the use of this drug. This is clearly a case of societal demand being both created and met by technology. But expediency has won out over reason, on both sides. Saturation advertising has created questionable personal priorities and has weakened the public concept of "fact," making the public susceptible to mass ripoffs. The so-called demands made on science by society are in fact demands created by corporate interests through the modulation and control of the collective will.

We are no longer in an era when practical applications of scientific research are unforeseeable and the human consequences unknown. Most of the science practiced today has at least a speculative relationship to a potential technology, and even when that is not true, we know enough about the relationship of science to technology and technology to society to know that caution is advisable. It has become the moral responsibility of scientists to consider the social implications of what they are doing—partly because they are inevitably the first to sense the approach of new technological capacities, and partly because there is no one else ready to take the burden of responsibility from them. If scientists will not accept this burden, then sooner or later society will decide to control the scientists.

6
Rousseau Revisited

The old relationship of mutual support and benefit between science and society is disintegrating. The reason for this is not hard to find: science's product is now so pervasive that society has been transformed by it into something radically different from what it was. A simplistic arrangement of producer and consumer between science and society is out of the question. The social, political, and economic elements of a technological society inevitably impinge on one another, often creating spurious needs and inane desires; it is no wonder that a chaotic mass of products and services is the result. Each useful item inspires the production of a host of similar items no more useful than the first and often less so, but we are expected to believe that the morass provides a wider choice: witness the number of brands of toothpaste distinguishable only by the amount spent on their promotion. When a useful technical innovation comes along, it is frequently distorted to fit meaningless needs created by extraneous forces: witness television with its crime and violence, punctuated by commercials. In order to form a more meaningful relationship between society and science-based technology, a new element of social consciousness is needed. A look at some recent events in the worlds of science and industry will help to bring this into focus.

The human virus called adenovirus-2 is relatively harmless

to humans; it usually causes symptoms resembling those of the common cold. Most people have been infected by it. Parts of this virus have been combined with parts of a monkey virus known as SV40 in order to construct a series of hybrid viruses that are useful in virus research. The monkey virus is able to transform normal human cells in the laboratory into typical cancer cells, but its effect on humans is unknown. The effect of the hybrid viruses on humans is also unknown, but because they may combine the ability to infect humans easily with the ability to cause cancer, they are potentially dangerous. The National Institutes of Health has alerted the scientific community to this possibility.

In 1971 Dr. Andrew M. Lewis, Jr., assumed the responsibility of distributing these hybrid viruses to research laboratories throughout the country (and perhaps abroad), for it is common practice to provide new and rare research materials to scientific colleagues. Because of the potential danger, Lewis felt uneasy about distributing the viruses and asked the recipient investigators to sign a Memorandum of Understanding and Agreement indicating that the researchers appreciated the hazard and would exercise due care. Before long, Lewis learned that the agreement was not being honored. Distraught, he stopped distributing the virus. He also reported his experience at the Asilomar Conference in 1975, where safety guidelines for recombinant DNA research were first drawn up: "The unwillingness of laboratory directors and interested investigators to respond to well-founded concerns and to accept responsibility for containing potentially hazardous agents would appear to have significant implications for any attempt to deal with the problems posed by bacterial plasmid recombination."[1] This statement was not warmly received by the molecular biologists present at Asilomar, who were in the process of making guidelines for themselves and did not want to hear that self-regulation might not work.

It was not very long before Lewis's apprehensions began to materialize. The incident involved the insertion and growth in bacteria of the gene for insulin, which was one of the potential applications of recombinant DNA technology discussed at Asilomar in 1975. The work was funded by the National Institutes of Health and carried out at the University of California at San Francisco during 1976–77.[2] Its success caused a great deal of excitement among molecular biologists because it was the first step leading toward a practical application of the new recombinant techniques: the production of insulin, a mammalian hormone, in bacteria. The work also caused concern in the science community and in Congress because some of the experiments were carried out with a plasmid vector that had not been certified as safe by the National Institutes of Health.

In the Senate hearings on the safety of recombinant DNA research,[3] the responsible scientists implied that there had been a misunderstanding concerning details of the mechanism for approval of the plasmid. Yet, the NIH guidelines clearly state that new plasmid vectors must be certified by the director of the National Institutes of Health before use. Apparently the scientists simply assumed that the plasmid would be automatically certified, and went ahead with their work. They stated at the hearings that they later destroyed the recombinant plasmids when they realized that the guidelines had been breached. There was considerable confusion in the testimony about the timing of that realization, however. In any event the work was suspended March 3, 1977, but the plasmids were not destroyed until at least a week or so later. The scientists claimed that they delayed destroying the plasmids because they were still expecting imminent certification from the National Institutes of Health. When it did not come, they finally destroyed the recombinant plasmid clones.

The following excerpt from the transcript of the hearings gives an indication of the confusion, if not duplicity, surrounding the incident:

Senator Stevenson: . . . Now my first question is to you, Dr. Boyer. Did you, as Dr. Rutter's memorandum so states, know—and on February 4 inform others—that pBR322 was not yet a certified vector?

Certainly, you don't have to look to Dr. Rutter.

Dr. Boyer: I didn't know whether he was going to answer it.

Senator Stevenson: I asked that question of you.

Dr. Boyer: OK. At that meeting we told members of Dr.

Goodman's laboratory and members of my own laboratory who were involved with various types of recombinant DNA experiments, either dealing with the P-3 laboratory, or not dealing with the P-3 laboratory, we informed them about procedural matters. We wanted to reiterate the procedural matters for the lab room and institute the use of a logbook for documenting activities in the P-3 laboratory.

At that time it was stated that the pBR322 plasmid, although approved, had not been officially sanctioned by NIH, by a written statement.

Senator Stevenson: So you did know at that point that it had not been certified; is that correct?

Dr. Boyer: I knew personally; yes.

Senator Stevenson: You said earlier, on January 16 you were told by Dr. Gartland that it had been approved subject to additional data. On February 4, you are stating to others that it had not been certified. What happened between those two dates? In those two weeks? You discovered at some point in between that it had not been certified, or did you know on January 16 that it had not been certified?

Dr. Boyer: I can't say for sure that I really understood the difference between certification and approval on January 16. It was shortly thereafter that I became aware of this difference in terminology, that there was approval at committee level and certification by the director.[4]

The senators were disturbed not only about the time element but also by the laxity and lack of authority at the university. When the breach of the guidelines had become evident there, the university chose not to notify the National Institutes of Health. Dr. Rutter stated in his testimony:

We felt that directly informing the National Institutes of Health would inevitably lead to public disclosure and debate about this incident and that would exacerabate the whole situation. . . . In the end, we chose not to inform the National Institutes of Health formally but to take the most conservative approach to the experiments themselves. We destroyed the clones.

In this case there was no question of a coverup. More

than 25 individuals in the department knew directly of the experiments themselves. We did not try to contain that information.[5]

In addressing Dr. Rutter, Senator Schmitt noted:

You say there was no coverup; but you did make a conscious decision not to inform the National Institutes of Health. . . . I think the rationale that some knowledge in the hands of the public could be dangerous or counterproductive is not really an acceptable rationale, even though I will admit that sometimes a little information can cause great misunderstandings. . . . But making an independent decision on what the public should and should not know is, I think counterproductive.[6]

At one point in the hearings, Senator Stevenson, exasperated, asked: "Would reasonable men, let alone scientists, have proceeded without some confirmation, without something in writing, without something that wouldn't put you in this preposterous position today of trying to reconstruct all of this? You say you don't want legislation. If there is legislation, you gentlemen would be the authors of it. I cannot, for the life of me, understand how reasonable men could have relied on rumors."

In an exchange of letters with Senator Stevenson, subsequently read into the Congressional Record, Wacław Szybalski of the University of Wisconsin came to the defense of Drs. Rutter, Boyer and Goodman. Professor Szybalski wrote:

Diabetes is a serious life-threatening illness causing human suffering.

The work of Rutter's, Boyer's and Goodman's teams offers hope of producing a cheap and abundant supply of human insulin, much better and safer than the present porcine insulin. . . .

How could a scientist, who knows that he might be able to save human lives and prevent human misery, delay his research just to avoid some unjustified criticism? I would not have much respect for a scientist who so worries about himself that he is not willing to help suffering humanity, especially when he knows that he could do it.

I would expect you, Senators Stevenson and Schmitt, to commend Drs. Rutter, Goodman and Boyer for trying to rapidly develop human insulin for the suffering and poor throughout the world.

To which Senator Stevenson replied:

The point, however, is that the scientific community has the responsibility to follow whatever guidelines are in effect until the rules are changed in an appropriate fashion. I am frankly surprised at your suggestion that Senators should commend scientists for violating these regulations simply because these scientists were engaged in important and useful work.

In his book *Recombinant DNA*, John Lear notes that this testimony is

a sufficiently sad account of disrespect for the democratic process. But the letters, memos, and other documents introduced into the official transcript during the four months following the end of the hearings show that virtually no one told the whole truth on the witness stand.[7]

The insulin incident, behind which the visible commercial application of the research looms large, shows how Big Science has taken on many attributes of big business, in its practical behavior as well as its aims. There is nothing new, however, in the competitive drive of the scientists. In a scholarly and often entertaining essay, the eminent sociologist Robert Merton documents numerous other cases as long ago as Sir Isaac Newton, demonstrating beyond the slightest doubt that rivalry for priority of discovery is foremost in the minds of many scientists and is exquisitely painful for the participants of the drama.[8]

In December 1977 a professor of biological chemistry at

Harvard Medical School was instructed by the National Institutes of Health to halt his research involving recombinant DNA technology. This action resulted from the discovery that Harvard had not filed the Memorandum of Understanding and Agreement required by the National Institutes of Health from laboratories engaged in recombinant DNA research. Officials at Harvard apparently did not know that one of their professors was carrying out experiments that should have been done in high-security laboratory facilities, which Harvard did not have. The original grant application, which was approved and funded by the National Institutes of Health, did not mention recombinant DNA; the situation was brought to light by the Environmental Defense Fund, which acquired its information through the Freedom of Information Act. In his defense, the professor asserted that the research "had been done in strict accord with National Institutes of Health guidelines."[9]

In an extremely clever allegory, Nicholas Wade of *Science* magazine satirized the attitude of many scientists toward the NIH guidelines, using Harvard as a focal point.[10] The article is in the form of a schoolboy's letter to President Carter. The letter starts: "The name of my school is the Harvard Medical School. You may have read in the newspapers that there was a boy here called Charlie Thomas who used to do experiments with recombinant DNA during biology class and who was told to stop them last December by the grownups at the National Institutes of Health." The letter continues: ". . . it wasn't our school's fault or anyone's fault in particular, it was just a general fault, all spread about among everybody involved like raspberry jam at a tea-party, and in any case the rules were too complicated even for very clever boys like us to understand." The letter concludes: "President Carter, even though the rules weren't exactly followed, nobody could have got hurt in any way by what happened. It's just that the rules are very new and complicated for us to understand, even though we are the cleverest boys probably in the whole country. Anyhow, we promise it will never happen again and we will be on our best behavior and work very hard in biology class and do great things like curing cancer, when we grow up."

The National Institutes of Health guidelines for recombinant DNA research were drawn up in 1976 by a committee of scientists in order to minimize the risks which they themselves had assessed. One member of the committee was Dr. Charles Thomas of Harvard.

Although there is no way of knowing how many similar infractions may be taking place in other laboratories, I believe that incidents like these are still rare. The attitudes that engendered them are not uncommon, however, and the situation is not likely to improve when familiarity has further dulled the investigators' appreciation of the need for caution.

In the field of biological research, the very recent development of recombinant DNA and other techniques related to genetic engineering has, for the first time, laid a *direct* responsibility on the research scientist in relation to the public welfare. There have been other, perhaps more overtly dangerous, areas of research (e.g., with contagious-disease-causing agents), but the concern of the scientist in those areas has been first of all for his own safety. Biological scientists have been largely free of a wider responsibility for public safety until now, and they are reluctant to give up that freedom. Still, they cannot resist grasping each new discovery and pursuing the knowledge it can bring, like Eve with the apple.

Scientists like to think of the scientific community as occupying a special position in society—disinterested, committed only to truth, and somehow therefore a social benefactor. The new relationship of biological science to the public brings into focus a missing element in this picture: the lack of a real social philosophy of science. The vague and permissive conviction that new knowledge is good for society is not sufficient, as many post-nuclear physicists would agree. Now is the time for professors and students to give serious thought to a research ethic similar to the medical ethic, one that would define the specific relationship of scientific research to individuals and society. The direct relationship of the laboratory to the public safety is only one aspect of the matter; the direction and aims of research, and the utilization of its fruits, are also on the scientist's conscience. These matters are now left largely to chance, or to private determination, by most scientists. Yet without defined and universally held principles of responsibility that reach beyond the laboratory to include all the ways that science impinges on society, the scientific community is just a self-interest group like any other, with its own delimited interests, its own motivations, its own rewards in terms of power, influence, and recognition, and its own ties to the Establishment and the status quo. Is science intrinsically different from industry in this respect? Or in its social conscience?

Let us take a look at the question of social responsibility in the field of applied technology.

Leptophos, or phosvel, is a pesticide that was manufactured by the Velsicol Corporation until January 1976. The pesticide causes paralysis, impotence, confusion, lethargy, severe lack of coordination, weakness, sweating, difficulty in swallowing, and vomiting. Production was stopped only after innumerable warnings from the Environmental Protection Agency and National Institute of Occupational Safety and Health and the poisoning of many workers at the Leptophos plant in Bayport, Texas.

The World Health Organization stated as early as November 1971 that "... Leptophos was shown to give neurotoxic effects ... this compound should not be used in [insect] vector control work and therefore further evaluation in the WHO Program should not be carried out." In 1973, a report of work done at Alexandria University in Egypt concluded that as a result of neurotoxic symptoms the pesticide "requires careful consideration before it is allowed to be freely used." Investigations at an EPA laboratory in the same year confirmed that the pesticide produced neurotoxic symptoms. However, disregarding these data and relying on information from the Velsicol Corporation which indicated Leptophos was safe, the EPA agreed to tolerate residues of the pesticide in imported vegetables.[11]

Since the controversy over Leptophos was persistent, the Environmental Protection Agency did not register the chemical, which would have permitted its use within the United States. The pesticide was used in Mexico, however, and contaminated tomatoes and lettuce imported from that country were sold on the American market. After further testing by the EPA confirmed the neurotoxic effects of Leptophos residues, the agency proposed to revoke the import tolerance in May 1975. Velsicol vigorously opposed this action and demanded the formation of a scientific advisory committee, as provided by law, to look into the situation. Velsicol did this in the face of a report from its own medical consultant, who stated in June 1975: ". . . there have been a series of unusual central nervous system illnesses. . . . I advise the company seriously consider halting manufacture of Leptophos until these matters are clarified."

Several months later, in October 1975, another medical consultant recommended that Velsicol "stop production of the material or store it until [Velsicol] can better determine complete implication of its use." In August 1976 Velsicol withdrew its application for registration of Leptophos, after the National Institute of Occupational Safety and Health inspectors threatened to visit the Velsicol plant. Velsicol decided to halt production of Leptophos in January 1976. The damage had already been done, however. Velsicol had sold the pesticide to fifty countries since 1971, and to twenty-nine countries in 1976. The United States imported 624 million pounds of Leptophos-treated tomatoes in 1976. In addition, the United States had been importing Leptophos-treated beans, peppers, cucumbers, peas, cantaloupes, eggplant, and squash since 1972. (Mexico finally banned the pesticide in April 1976.) The Food and Drug Administration, which is responsible for monitoring pesticide residues on such imports, was able to sample only 650 out of 35,668 shipments in 1976, after the residue tolerance had been revoked. Have you ever had any symptoms that were hard to explain?

It is obvious from the data outlined above, given in detail in a congressional staff report, that the Velsicol Corporation had withheld vital information from the EPA and had continued to manufacture a product it knew to be dangerous.[12] The reason seems clear: the U.S. market for pesticides was $1.9 billion in 1974 and may be as high as $3.3 billion in 1984.

The case of Dioxin illustrates a different kind of industrial irresponsibility: the accidental spread of a toxic substance in the environment, followed by a large-scale coverup.[13] [14] Dioxin is a by-product in the manufacture of tricholorophenol, which is used in the manufacture of a herbicide. On July 10, 1976, an explosion in a chemical factory at Seveso, Italy, resulted in a spray that contaminated the surrounding area with Dioxin, a highly toxic chemical. Two days later, Givaudan ICMESA, a subsidiary of Hoffman-La Roche, reported the accident to local authorities in Seveso, who in turn announced that consumption of food crops should cease. On July 15, animals began to die; about a week later, townspeople began to exhibit rashes and burns. Local citizens were told that there was no need to panic. Because the situation showed no signs of improvement, however, evacuation of the residents of Seveso began about three weeks after the explosion.

This slow reaction to the catastrophe was no doubt based on the implication by ICMESA that the escaped vapor contained only trichlorophenol, a much less toxic substance than Dioxin. The firm knew, however, that Dioxin was a by-product in the manufacture of trichlorophenol. Local authorities at Seveso were unaware of previous accidents in the manufacture of this chemical: at Monsanto (U.S.) in 1949, Badische Anilin und Soda Fabrik AG (West Germany) in 1953, Dow Chemical (U.S.) in the 1960s, Philips Duphar (Netherlands) in 1963; and Coalite and Chemical Products (UK) in 1968.[15] Had they known of these incidents, they would have reacted instantly and ordered evacuation and medical care.

There were other signs that ICMESA was apprehensive about revealing details of the incident. Local authorities at Seveso tried to learn the nature of the manufacturing process, but they were told nothing for some time. Finally, according to *Nature*,[16] ICMESA came through with half-truths; they did not mention Dioxin as a serious by-product. Workers and local residents were kept in ignorance of the real dangers and the extent of the contamination of the soil and countryside. At last an Italian commission was set up, and the truth became known. Two top officials of ICMESA were arrested and charged with culpably causing a disaster, and civil proceedings were brought against Hoffman-La Roche by Italian authorities.[17] Later the American environmental scientists

Barry Commoner and Robert Scott informed the commission of the results of U.S. Air Force studies carried out with Dioxin in connection with the use of herbicides ("Agent Orange") in Vietnam. The evidence allowed them to calculate that it would take fourteen years for Dioxin to break down to undetectable levels in the soil.[18]

The clinical symptoms for Dioxin poisoning, for which there is no antidote, have been known for some time. They include atrophy and necrosis of internal organs as well as the very disfiguring chloracne, a severe acne condition caused by chloro-organic compounds. In addition, Dioxin has been shown to cause birth defects in mice. When they learned of this, 730 pregnant women of Seveso applied for legal abortions for fear of giving birth to children with congenital abnormalities.

The persistence of Dioxin was also well known before the accident at Seveso. After the explosion at Philips Duphar in the Netherlands, for example, the walls could not be successfully decontaminated; it was necessary to dismantle the building brick by brick and imbed the rubble in concrete containers that were then dumped into the Atlantic.

It is hard to imagine what rational motive prompted the reaction of ICMESA to the disaster. It was as if by closing their eyes to the situation, it would go away. Yet this is not an unusual response to technological problems in our large-scale technocracy, and it is not confined to the industrial sector. We have made a machine that eludes human control in many ways, and the psychological response is to relinquish all efforts to guide it. This is strikingly documented by an almost incredible series of events that occurred in Michigan a few years ago, which were characterized by the adamant refusal of persons at every level to perceive danger, use judgment, or exercise responsibility.

In 1973, some commercial cattle feed used by Michigan farmers was poisoned inadvertently by the inclusion of a flame retardant, PBB (polybrominated biphenyls), rather than the usual food supplement. The problem originated at the Michigan Chemical Corporation, which manufactures both flame retardants and feed supplements and apparently confused them when filling an order from the Farm Bureau. The initial error was caused by irresponsibility on the part of workers, and the failure of the management to inform them of the hazardous nature of the material they were dispensing.[19] The Farm Bureau, a cooperative that formulates cattle feed, compounded the error by ignoring the PBB labels on the material and using it for mixing in the feed. The poisoned feed was then distributed.

After noting the obvious symptoms of sick calves, the farmers sought help from the Michigan Bureau of Agriculture. They received only token cooperation and were forced to investigate the matter at their own expense. Even after the poison had been identified as PBB in the feed, farmers were not warned of it. A University of Michigan scientist tried to alert state authorities to evidence that PBB causes birth defects and might pose a human health hazard, but no one listened.

Many farmers and their families became ill, and the cattle developed weird symptoms. Because the Michigan Department of Health could not diagnose the problem, sick animals and their products were allowed to be sold for nine months. As a result, most people in Michigan have measurable levels of PBB in their bodies. All during this period, tens of thousands of sick animals had to be destroyed and buried. Some farmers with so-called high-level cattle were compensated; farmers with lower-level cattle were not. This action was completely arbitrary because the meaning of PBB levels was unknown.

As late as March 1975, the Michigan Department of Public Health insisted that there was no pattern of human illness due to PBB. The assertion was restated by the state's director of health. On television the community health chief emphasized that there was no health problem.[20] Even after PBB was found in nursing mothers, the state health director kept assuring everyone that there were no sick babies. As early as October 1974, scientists at Mt. Sinai Hospital in New York offered to investigate the problem free of charge, but for nineteen months Michigan's governor refused their aid, until circumstances literally forced him to accept the offer. The Mt. Sinai report, issued in January 1976, indicated that there was indeed a serious problem. Continued university studies on the effects and levels of PBB, financed by the farmers themselves, have shown that human exposure is widespread and that PBB at very low levels has the potential to cause cancer and birth defects in humans, in addition to the many symptoms already visible.

It is of interest that the Michigan Chemical Corporation is part of the Velsicol Corporation, which in turn is part of a Chicago-based conglomerate. Velsicol is known for its high profits, for the manufacture of Tris BP, a carcinogenic flame retardant used to treat babies' clothing, and for the Leptophos fiasco. The Michigan Chemical Corporation operates in a ramshackle factory, polluting the soil and rivers, according to state records.[21] It was alleged to have sent only a single ton of PBB to the Farm Bureau for mixing in the feed, but this is uncertain; nineteen tons of PBB could not be accounted for by the company. The company also had told its workers that the material was safe and could even be eaten.

The collateral deceits were the least of the problem, however. First and foremost, the disaster was characterized by ir- responsibility all along the line, from the individual workman mixing feed up to the governor. The evidence of carelessness, not to say coverup, by various agencies of the state is overwhelming and appalling.[22] For the first two years of the disaster, the governor concurred in the decisions and actions of the Departments of Agriculture and Health. Even after the governor was pressured into appointing an expert panel to study the PBB situation, the State Agriculture Commission refused to follow the panel's recommendations and the gov- ernor did not override the commission's decision. It has been asserted by a former member of the Michigan House of Rep- resentatives that the various state agencies were involved in a coverup.[23]

The sequence of events, starting with the initial misreading of the labels on the PBB containers and through to the final investigations, follows the familiar pattern of an inevitable chain reaction brought about by *de facto* irresponsibility on the part of all involved. The series of actions was neither designed nor intended; what is worse, none of the actions deviated sig- nificantly from what has come to be regarded as normal behavior, up to and including the apparent coverup. This disaster is only one of a class with which we are all only too familiar. It represents a symptom of ultratechnology: no individual feels obliged to intervene in the inexorable progress of technology, not even when human health and lives are at stake. It is as if the end result were preordained; no one or nothing could alter the course of events set by the initial act. These are symptoms of deep frustration and a consuming hopelessness buried in our subconscious minds, implying a helplessness so profound as to preclude a call for help.

Technology is based on science; and technology is not fulfilling its obligations to society. The few examples just offered are symptomatic, if not typical, of a widespread malaise in the relationship of modern technology to human beings and their environment. This fact changes the nature of responsibility all along the line. In an ideal world, scientists could perhaps justifiably pursue any and all knowledge, leaving it to others to evaluate and utilize their results for the common good. But when it is apparent that those farther along in the chain cannot be counted on to exercise social responsibility, it becomes irresponsible for scientists—the prime movers in the technological process—to continue the indiscriminate fueling of technology. It is now the scientist's duty, by default, to direct the limited resources of research away from goals that could easily be abused (however potentially useful they may be) and toward those that are most likely, given the realities of the system, to be utilized for the public benefit. This means that the scientist should live up to his title as Doctor of Philosophy, broaden his learning and his understanding, and give serious thought to the major concerns of humanity and the ways in which science may—or may not—be able to contribute to them. The scientist has an obligation to act politically to combat inappropriate applications of science, whether these are harmful, useless, or diverting resources from more significant pursuits. He should participate in alerting the public to any potential problems; the scientist is, after all, in the best possible position to recognize them. And he should direct his own research accordingly. This is the scientist's responsibility in the world we live in, and it is not a bit more utopian than the prevailing contention that social responsibility belongs not to the scientist but to the technologist.

Every now and then, something encouragingly responsible takes place in the scientific world. The first stirrings of conscience over the hazards of recombinant DNA for instance. Or the singularly excellent report of the Select Committee on GRAS Substances,[24] a scientific evaluation of industrial usage that discusses candidly the goals, shortcomings, and strengths of the endeavor. The report was made by a disinterested group, and a special effort was made to eliminate bias. This is in sharp contrast to so many investigations in which committees—even those that originate in the loftiest places, such as the National Academy of Sciences or the White House—are unable to free themselves from Establishment views or from political, commercial, or self-interests.

GRAS is an acronym for Generally Recognized As Safe; it refers to food ingredients. The Select Committee was organized in 1972 by the Life Sciences Research Office of the Federation of American Societies for Experimental Biology. The committee issued its report in 1977 after a five-year study involving fifty executive sessions. Its stated purposes were fourfold:

1. To illustrate the range of factors to be taken into consideration in the safety assessment of a given food ingredient
2. To offer estimates on the state of the art and commentaries on the nature of technical dilemmas encountered in rendering scientific judgments on food safety
3. To provide suggestions concerning the philosophical, procedural, and scientific ramifications of the evaluation process
4. To point out needed research to improve the validity and meaningfulness of the associated data

Serious problems in evaluating the data were encountered in working toward these goals, but in spite of unavoidable shortcomings the study represents an important contribution to public and scientific dialogue. All details of the committee's structure and operation received considerable attention. For example, its membership was broad—it represented ten fields of endeavor; the orientation of each member was considered so that there was no apparent imbalance resulting from a skewed polarization of viewpoints. As the committee put it: "We preferred the approach of collective reasoning among members with a minimum of bias in either direction." The committee also emphasized the need for public participation, which was formally requested through the *Federal Register*, a legal but not very "public" medium. (In my view this is an unsatisfactory means of recruiting public input and represents a weakness in the committee's procedures.)

The committee went into considerable detail on all aspects of food additives, starting with the collection of relevant data. It studied a wide range of subjects, such as the effects of additives on fetuses and neonates; the carcinogenicity, mutagenicity, and allergenicity of additives, and also their interaction with drugs. In brief, the study was comprehensive. Because myriad variables could not be quantified, the committee chose to use "reasoned judgment" in arriving at the hazards of each additive. The deliberations produced a logical end point, namely, a consensus opinion, favoring the alternative that would produce the least harm. From a practical point of view, the committee stated that there was no satisfactory method for calculating a risk-benefit ratio. It concluded with a series of suggestions detailing a program for evaluating food additives from the point of view of demonstrable safety rather than lack of a demonstrable hazard.

The report of the Select Committee on GRAS Substances is unique in several respects. For one thing, the report makes absolutely clear that the effects of substances ingested by humans, be they food additives or drugs, are impossible to quantify. The presence of "U.S. certified food color" on a label takes on a different meaning after reading the GRAS report. The committee's philosophic reference, indeed its human approach, is evident throughout the report; these scientists discovered that science is weak in the area of evaluation of human hazards, and they imply that scientific humility should enter the considerations. The report explicitly recognizes that value judgments will always be a part of the reasoning process in risk-benefit analyses. It is refreshing to find that some scientists realize that subjectivity has its place, even in matters of science.

The GRAS report is a signal effort, embodying a sound and humane approach, in sharp contrast to the usual *modus operandi* in which scientists ignore the fate of their findings while industry applies them in ways that are often irresponsible. The Federation of American Societies for Experimental Biology submitted the report to the Food and Drug Administration, where it is being used as a basis for the regulation of commercial use of the food additives studied.

This example gives a hint of what might be done to influence the direction of technology if the scientific community *as a whole* shook off its apathy and took on an active social conscience. Prestigious studies like the GRAS report, relating to current or proposed technological practices, would be hard for government and industry to ignore, particularly if they were carried out with the knowledge and participation of the public. An active and concerted effort could also be organized to identify the more socially useful directions for research and technological development. A new sense of awareness and a wholeness could arise out of the fragmented scientific world if the scientific community would only recognize that there is an urgent mission to be shared: to rescue society from the collision course on which it has been set by the blind progress of technology.

7
From Truth to Power

DNA has become a household word. It symbolizes scientific accomplishments that tell us in concrete terms about the laws of heredity. DNA represents a basic scientific truth that will remain with us forever. Just a few decades ago, no one could have imagined that the ordering of the chemical components in DNA could determine all the genetic traits of living organisms. The discovery of this phenomenon has to be an utterly exhilarating experience for all who learn of it. For the scientist, it is this kind of emotional experience, placed in an intellectual context, that produces an inexorable drive for further experimentation and causes him to submerge himself deeper into "truth." At the same time, without his intending it, this process generates elitism and the drive for power, individual as well as collective.

The biomedical research community has provided one of the most illuminating case histories in recent times of the seductions of power along the path to truth. I refer to the circumstances surrounding the development of recombinant DNA techniques, which enable scientists to manipulate inheritance by transferring genetic material between organisms. The whole drama started when several scientists expressed concern over the safety of certain recombinant DNA experiments involving the use of cancer viruses.[1] After many telephone calls and private conferences among colleagues interested in this

new technology, an international scientific conference was organized at Asilomar to consider the biohazards associated
with recombinant DNA. These discussions, which eventually
formed the basis of the National Institutes of Health
guidelines, adhered closely to the limited facts at hand. A natural but unfortunate consequence was that the resulting
guidelines were concerned primarily with those experiments
with which the Asilomar molecular biologists were most familiar—their own. When it came time to consider national safety
legislation, in 1977, the molecular biologists realized that they
had trapped themselves: the concern of the public had also
been focused on safety measures for their research. The safety
of industrial processes using recombinant DNA, and most importantly, the long-range consequences of the technology, had
been neglected. The scientists set about to "reorient" senators
and congressmen so that legal restrictions on their research
would be minimal or nonexistent. They went about this by
culling facts from the existing literature as well as using new
and fragmentary data, most of which had not yet been published in standard scientific journals, to make a case for the
safety of current research. Much information passed by word
of mouth, and more through the lay press; both channels
transmitted the facts in incomplete and sometimes distorted
form. But the legislators had only these data with which to
work.

The job of the professional politician is to seek compromise
in order to minimize conflicts, thereby permitting accom-
modation of the maximum number of people.[2] The scientist,
on the other hand, is not expected to seek compromises; he is
to solve problems on the basis of demonstrable hard facts. He
has no right to bend the truth to achieve a more desirable
outcome, even if that seems to be the only way out. But many
scientists became "politicians" in the recombinant DNA controversy. Without due consideration of the fundamental issues, they manipulated scientific information in order to
achieve certain political goals of the scientific community. I do
not mean that scientists have an unusual thirst for power; in
general this is not true. But, unlike other pressure groups, scientists as a body continue to believe that they are purely ob-

jective and disinterested even when they have left the ivory tower far behind. They are reluctant to recognize that modern science is continually in the political arena, both influenced by, and influencing, the world outside. In chapter 5 I examined the external circumstances that mold Big Science, and in this chapter I show how these forces create pressures that drive the scientist down the path from truth to power.

Before 1944, when DNA was shown to be the genetic chemical, it had been widely assumed that proteins alone comprise the important macromolecules of living cells, carrying out vital functions as well as determining hereditary traits. Although many chemical investigations had been carried out on DNA both before and immediately after 1944, it was not until almost a decade later that the DNA structural hypothesis of Watson and Crick provided the basis for the study of molecular genetics, one of the strongholds of molecular biology.[3] But Watson and Crick did not create their hypothesis out of the blue; it was based on the painstaking chemical work of Chargaff and his collaborators, and on the X-ray crystallographic studies of Franklin and Wilkins.[4] Without their work, the Watson-Crick hypothesis could not have materialized.[5] In addition to these two major experimental contributions, scores of other chemical facts entered into the proposed structure.[6]

All told, many man-years of research were involved. At the time much of the work was done it had no apparent relationship to what finally emerged as the actual structure and function of DNA; it was work done for the sheer intellectual satisfaction of it. There was no general public from whom to expect acclaim; there were literally only a few score of other interested scientists. This was not a unique situation in the days of Little Science. In modest and sparsely equipped laboratories, with instruments that were primitive by today's standards, an intimate and intense scientific intercourse took place among the small number of workers. Relatively few papers were published, but each was packed with essential scientific data, not skimpy and difficult to assess or repeat, as is often the case today. The quality of Little Science was not inferior to today's science, in spite of limited resources; in many instances the quality was higher than it is now. Little Science was an activity more closely approximating the ideal for the pursuit of pure knowledge. The search was in large part unencumbered by extraneous elements. This was the almost serene atmosphere in which DNA was born. Herein are a few of my thoughts on the unrecorded early history and evolution of DNA research.

The double helical structure of DNA requires specific physical-chemical forces to hold the helices together. These forces arise from complementary interactions between chemical groups (purine and pyrimidine bases) in the DNA. The interactions produce base-pairing. Without base-pairing, DNA would be a formless coil instead of a majestic, erect helix —one of many beautifully symmetric mathematical curves. I might add in passing that helical structures in biological systems were first observed by Linus Pauling; doubtless this work had a profound influence on Watson and Crick.

The base-pairing hypothesis of Watson and Crick was a stroke of genius. The pairing of the base adenine to the base thymine, and similarly of guanine to cytosine, allowed for a regular double helical structure and at the same time crystallized the concept of complementarity in biological systems. Complementarity (discussed in chapter 3) is the chemical basis for the observation that like begets like; it explains how hereditary traits persist from one generation to the next. Interestingly, Niels Bohr, the theoretical physicist, had proposed the principle of complementarity in physical systems in the early 1930s, and Max Delbrück and Linus Pauling had discussed the need for it in biological systems in the 1940s.[7] For different reasons, these scientists had seen the need for the concept of complementarity, a thread connecting the physical and biological worlds. But this notion was ahead of its time, in those early years, and as often happens in science, the concept lay fallow. Only the convergence of the relevant chemical and physical data assembled at the proper moment led Watson and Crick to their structure. There is no question in anyone's mind that the hypothesis for DNA structure and function represents one of the great intellectual constructs of the twentieth century. This was the dawn of molecular biology. A new dimension had been added to biological "truth."

To this day I am still amazed that Watson and Crick foresaw that their proposed structure posed what seemed to be insurmountable difficulties for the replication mechanism of DNA, yet they would not abandon the structure. Crick would often say that he thought the structure was "essentially correct," in spite of some "nasty structural details." I can still see him rocking in a swivel chair during a visit to Sloan-Kettering Institute in 1954, insisting: "It [the structure] just has to be right; it has the ring of truth." Crick is an eloquent speaker; his excitement is infectious whether he is speaking to an individual or a group. We were all soon convinced of the essential correctness of the structure. On the assumption that the structure of DNA had been solved in 1953, Crick began developing other theories for the structure and function of macromolecules, especially RNA, which is chemically related to DNA. He was sure that RNA molecules carried the "message" encoded in DNA and were directly involved in the synthesis of proteins. Crick and others, such as George Gamow, the theoretical physicist, soon developed the idea of the genetic code, which related the chemical composition of the DNA gene to that of the protein it specified.[8] During the next several years, an ever-increasing array of molecular biological discoveries and theories were discussed at symposiums and private conferences, and global thinking became the order of the day. Its hallmark is the unified theory, embracing all facts and accounting for them in a coherent framework that can then be used to propose future experiments. This is the pursuit of truth at its most intense.

A significant portion of these "think-tank" activities occurred at the famous Gordon Research Conferences, which take place every summer in New Hampton, New Hampshire. Amid quiet, pleasant surroundings in the foothills of the White Mountains, about one hundred nucleic acid researchers would gather each year for a full week of scientific intercourse. From morning until late at night, scientists would talk, drink beer, and listen to the most recent results of molecular biology. The meetings were characterized by a free exchange among scientists, but the public and the press were completely excluded, and the proceedings were not published. An attempt by a *New York Times* reporter to crash the proceedings one year was aborted when the scientists voted unanimously to exclude him.

Each year, Francis Crick would attend the conference and fire up all nucleic aciders with his imaginative and usually correct hypotheses about DNA functions. It seemed as though the entire scientific community was at his disposal for him to direct their experiments from his podium; it was almost a kind of remote-control research. He was unique among the ringleaders, and I guess that his "sermons" encouraged the many hopeful global thinkers who were always trying to emulate him. Endless series of experiments and theories had their inception at New Hampton.

But the glow surrounding DNA cast bigger shadows than might have been anticipated. Its incipient power could be felt everywhere, as though science was gestating, ready to burst forth with new and more profound phenomena at any moment. As science probed more deeply into basic biological mechanisms, each successful discovery was accompanied by a feeling, growing stronger with time, that perhaps all of nature could be "explained" by the molecular biologist. The train of throught is understandable, for a great deal had been learned: the nature of the genetic material; important facts concerning how bacterial DNA replicates; the nature of some control elements; a fantastic variety of details regarding the synthesis of proteins, including the structure of the components involved; and, in latter days, the manipulation of genes through recombinant DNA technology, which allows scientists to put genes exactly where they want them. These are only a few of the more profound discoveries. The last item—the manipulation of genes—gives a clue to the power of molecular biology. It is in some ways a power over life. At the moment this power is being exercised on bacteria only, but little imagination is required to extrapolate to the use of this technology in other living systems.

During this period, many molecular biologists, assessing the discoveries and feeling quite confident about their new-found power, began more than ever to think along Cartesian lines, perhaps subconsciously. Descartes's aphorism, ". . . we can employ [forces] in all those uses to which they are adapted and thus render ourselves the masters and possessors of nature," seems to describe rather accurately a predominant attitude of a number of scientists, not only in the controversy over recombinant DNA, but in much broader areas. At scientific meetings on subjects like "Man and His Future" and "Science and Civilization," scientists discussed such subjects as the need for "a global evolutionary policy, to which we shall have to adjust our economic and social and national policies. . . . Eventually, the prospect of radical eugenic improvement could become one of the mainsprings of man's evolutionary advance." [9]

Dr. Philip Siekevitz of Rockefeller University remarked on this trend in 1970, pointing out "the remarkable hubris" of scientists:

> I think our greatest sin is to presume to know much more than we do, and even if we don't, we give the impression that we do, and so the world takes our tentative findings and makes them actualities. . . .[10]

It would be unthinkable and wrong to ascribe ulterior motives to Descartes, despite his pretentious attitude and aloofness (if not disdain) toward the public, and it would be equally unjust to entertain the notion that the modern scientist is scheming to rule his neighbors or conquer nature. I prefer to think of the modern scientist as an ordinary mortal having his own personal desires, dotted with altruistic aims, sometimes carried away by grandiose schemes and sometimes blind to the fact that unintended harm may accompany the application of his or her scientific knowledge.

The path from truth to power, at least in the early stages, is apt to be tortuous. A scientist's euphoric feelings about a discovery he has made, and his enjoyment of peer approval, impel him to investigate the subject more deeply; this may lead to the "need" for more staff, say, another postdoctoral fellow in the laboratory. Usually this can materialize only if the investigator can manage another grant. If he obtains additional funds, he then finds the need for more space and equipment. Soon there follows the need for promotion to a higher rank, which may be a prerequisite for further expansion of his research effort or nomination to an honorary society. And so the seed is sown; and it is nurtured by the system. But there is a price. Big Science, requiring, as it does, expanded resources, forces the investigator to compete for funds, which seem always to be in short supply. Grant applications become a primary and not a secondary occupation. To procure funds becomes a matter of management practices rather than science: How best to write a research proposal that will be funded? Can it be written in such a way that the investigator's scientific interests dovetail with currently popular mission-oriented aims? How best to acquire from the university or some other institution more laboratory space? These extraneous matters claim the scientist's attention and cause him to become a manipulator of programs, funds, and space. The Big Scientist is thus caught up in a power structure in which he is forced to participate if he wishes to remain viable. Concomitantly, the cherished activity—research in pure science—has been dealt a heavy blow.

For those who are not satisfied with mere survival and who wish to become leaders, one route is through the highly coveted Nobel Prize. To "win" the prize one needs, in addition to scientific achievement, a modicum of political maneuvering, and good public relations.[11] The scientist may start by moving to a well-known university with a bustling department filled with members of the National Academy of Sciences and a few Nobel laureates thrown in for good measure. Here the Big Scientist has all the opportunities for becoming a science potentate, for the power exercized by Nobel laureates and science academicians is legendary. The Big Scientist can now enter the politics of science. He and his colleagues will provide advice, solicited and proffered, individually or through committees, for government science policy decisions. These *de facto* leaders are therefore in a position to formulate policy for their colleagues, and what is more important, for the community at large. Their values and backgrounds become important, for there is no effective input from other quarters. Members of the public, who not only pick up the tab but are also the ones ultimately affected by decisions, do not usually have much opportunity to participate. I shall return to this subject later.

The controversy over recombinant DNA technology is a case in point concerning the vast influence of official scientific organizations, as well as of small but powerful cliques situated in prestigious universities. It also illustrates in a concrete way how scientific truth can be forced into strange configurations in the name of freedom of inquiry, and how the emotional appeal of the Nobel Prize and other amenities can help convert the pursuit of truth into the exercise of power.

The recombinant DNA debate made it plainly evident that this field had become politicized and that scientists had lost not only their scientific innocence but also their political naiveté. Concern over the hazards of this new technology was first voiced by Robert Pollack in June 1971. Professor Pollack pointed out the potential danger of some experiments being carried out by Paul Berg of Stanford University. After a private discussion, both men were desirous of responsible action; wheels were set in motion to look for solutions. But what started out as a laudable attempt at sober action soon became tainted. A brief recapitulation of the history leading to the defeat in 1977 of legislation for the regulation of recombinant DNA technology will demonstrate that the politicization of this branch of science was not a clean business.

Realizing the intrinsic hazards of the technology, the scientists involved were quick to alert other scientists to possible dangers.[12] The members of a Gordon Research Conference voted to request Dr. Philip Handler, president of the National Academy of Sciences, "to establish a study committee to consider [the recombinant DNA] problem and to recommend specific actions or guidelines, should that seem appropriate." Accordingly, the National Academy of Sciences formed a committee that recommended a partial moratorium, during which time two of the most hazardous types of experiment were not to be performed. This action was followed by the formation of the Recombinant DNA Advisory Committee, which was charged by the National Institutes of Health with formulating guidelines for recombinant DNA work. This committee called four meetings between February and December 1975, and sent a set of proposed guidelines to the NIH in January 1976.

In February a "public" hearing was sponsored by the NIH.[13] It should be emphasized that meetings legally become public when they are announced in the *Federal Register*. Needless to say, this type of notice does not reach the public directly or effectively. A number of public-interest organizations, including Friends of the Earth, the Sierra Club, Science for the People, the Environmental Defense Fund, the Coalition for Responsible Genetic Research, and the National Resources Defense Council, have registered dissatisfaction over the lack of effective and meaningful public input on this issue, in view of its relevance to public safety and environmental preservation. The Committee for Human Values of the National Conference of Catholic Bishops expressed concern and called for a pause for reflection.[14] Many individual scientists also expressed grave concern over the lack of participation by environmentalists, ethicists, public health officers, and representatives of the technicians who actually carry out recombinant DNA experiments.[15]

The completed guidelines were issued by the NIH in June 1976, and an environmental impact statement was published in October 1977. Although the guidelines became a *fait accompli* without significant public participation, there was considerable public interest in the matter of recombinant DNA technology. The issue was first publicized in the spring of 1975, when a controversy arose at the University of Michigan at Ann Arbor and committees were appointed to examine the advisability of constructing a high-security (P3) recombinant DNA laboratory on the campus.[16] Eventually, in March 1976, the university decided to proceed with the proposed construction of the P3 facility, despite opposing arguments courageously set forth by faculty members Susan Wright, Arthur Schwartz, and Shaw Livermore.[17]

The controversy at Michigan was just one sign of a rising opposition among scientists to the precipitous haste with which a small group was pushing ahead with the technology without waiting for adequate discussion and evaluation of its hazards. Several scientists took it upon themselves to circulate private letters of caution, and one announced that, as a member of a grant-reviewing committee, he would vote against the funding of research using the new techniques. The centers of scientific power reacted immediately with a campaign of covert intimidation, letting it be known that job security or scientific standing was at stake. Professor George Wald of Harvard noted in 1977:

> The conviction is widely distributed among young scientists and people about to get their degrees and nontenured young faculty that if one ever expects a job or if one ever is to expect support from the granting agencies, or continued support from NIH, it is best to shut up about this.[18]

Influential scientists with an interest in the technology began to recant their initial concerns and apply pressures to others to follow suit. The stirrings of conscience among the scientific community were effectively stamped out, to be replaced by a widespread fear of involvement in societal issues. This was a serious setback to the progress of social maturity among scientists. It also provided a sad example of the way in which a handful of power-oriented scientists can succeed in using the scientific community for its own purposes.

Without question the now-famous "Cambridge affair" was responsible for bringing the issues of recombinant DNA technology to the public eye in a most dramatic fashion. The furor was over the possible construction of a P3 recombinant DNA facility in Harvard's Biological Laboratories, a matter that was hotly discussed at faculty meetings even before it came to the attention of the mayor and citizens of Cambridge. Public fears were high, and the faculty was divided. As a result of public hearings held on June 23 and July 7, 1976, the Cambridge City Council voted for a three-month "good faith" moratorium on the construction of high-security research facilities. The council then set up a review board to decide whether the P3 facility should proceed. After many hours of hearings, the citizens' review board arrived at a decision in January 1977. The decision was important because it was the first time a public panel had entered meaningfully into the debate on recombinant DNA technology. The review board wisely noted in its report that "knowledge whether for its own sake or for its potential benefits to mankind cannot serve as a justification for introducing risks to the public unless an informed citizenry is willing to accept those risks.[19] The city council accepted the review board's recommendations in February 1977, permitting the P3 construction to proceed but imposing safety regulations stricter than the NIH guidelines.

Following the example of Cambridge, the New York State Attorney General's Environmental Health Bureau investigated the issue in October 1976.[20] After public hearings, a bill to regulate the safety of recombinant DNA activities was introduced and passed by the legislature, but it was vetoed by the governor after heavy pressure from a small group of scientists. The California legislature also held hearings, and introduced a bill in March 1977. The following month, Maryland enacted legislation extending the NIH guidelines to industry. Public action also took place in San Diego, California; Princeton, New Jersey; Bloomington, Indiana; Amherst, Massachusetts; and many other communities.

The public's interest in the recombinant DNA controversy was reflected in the U.S. Congress, where twelve regulatory bills were introduced in the House and Senate in 1977.[21] In early 1977, Senate Bill S.1217, a well-thought-out and comprehensive document, seemed likely to pass into law. This bill was drawn up by the Senate Subcommittee on Health, chaired by Edward Kennedy. It contained not only safety regulations but also provisions for public input and enforcement of the regulations. Opponents of S.1217 were highly critical of the proposed thirteen-member regulatory commission, which was to include both scientists and public representatives, and also of the provision that would permit local communities the choice of adopting regulations more stringent than the federal measures. The House measure (H.R.7897) drawn up by the Rogers Committee on Health and the Environment was also a serious contender. This was a far less restrictive document than the Kennedy bill, and was at first favored by proponents of recombinant DNA technology. However, by midsummer of 1977 the proponents had become still more uneasy about regulation, and they undertook a series of concerted actions to undermine all legislative efforts. This created an almost chaotic situation. Congressmen and the public were suddenly presented with a duststorm of "new facts" purporting to show that the risks of recombinant DNA research had been highly overestimated.

Four major events served to undercut the pending 1977 legislation and eventually also led to revision of the NIH guidelines to make them less restrictive. Two of the events resulted from the efforts of individual scientists, Roy Curtiss III and Stanley Cohen; the third event was a scientific conference, and the fourth was a massive campaign originating in the power structure of the science community.

Dr. Curtiss, a professor of microbiology at the University of Alabama Medical School, wrote a letter to D. S. Fredrickson, the director of NIH, stating his reasons for believing that there were minimal hazards in recombinant DNA techniques using certain strains of E. coli bacteria, and that restrictive legislation should be limited.[22] The letter was freely passed around and was influential in legislative circles. Briefly, the situation at that time was as follows. Dr. Curtiss had developed a highly enfeebled derivative of the bacterium E. coli K12, called χ 1776, which contains thirteen mutations. The mutations were designed to make it virtually impossible for the bacteria to survive except under special laboratory conditions. In his April 1977 letter, Dr. Curtiss discussed the probabilities of the transfer of recombinant DNA from an enfeebled bacterium to a normal one under various conditions. For the widely used but less drastically enfeebled strain E. coli K12, the probability of transfer of a (nonconjugative) plasmid was estimated to be 10^{-16} per surviving bacterium per day per intestine. For χ 1776 the estimated probability was lower. Curtiss also discussed the effect of recombinant DNA on the viability of bacteria containing such DNA. Finally he stated: "In summary, after pondering these and other types of errors [accidental events], I am convinced, because of the need for a sequence of errors and the improbabilities of constructing a microbe that both has a competitive advantage and displays a harmful trait, that construction and use of E. coli K12 strains with recombinant DNA poses no threat whatsoever to humans (or other organisms) except for the remote chance that an individual constructing or using such strains, as discussed above in the first examples of potential errors, might experience some ill effects." I have already indicated in chapter 3 that the results of actual experiments with E. coli, carried out recently for the purpose of risk assessment, have led Dr. Curtiss to alter his opinion on this score.

Dr. Curtiss and his team performed a heroic task in constructing the enfeebled E. coli strain X 1776. The genetic techniques were unquestioned and the bacteria were genuinely enfeebled, but in practice, as opposed to *a priori* theory, this is turning out to be an insufficient safeguard. Nonetheless, E. coli is still the major cloning host in laboratory use. Some scientists have argued that a bacterial species incapable of residing in humans would be a better choice than E. coli as a host for recombinant DNA.[23] Be that as it may, Curtiss's original arguments refer only to the safety of recombinant DNA in E. coli and its plasmids, and not to the different hazards posed by the variety of new hosts and vectors now coming into use. Overly sanguine as they were, Curtiss's probability calculations for E. coli (10^{-16}—10^{-20}) were still not sufficient to guarantee safety when coupled to a high-consequence situation.[24] As I pointed out in chapter 3, accidents with more accurately calculated probabilities of 10^{-18} and 10^{-20} have already occurred. While it is evident from Dr. Curtiss's letter that he did not intend to overstate his case, the letter was used to advantage by a number of influential lobbyists who may have been less scrupulous.

At about the same time, Dr. Stanley Cohen, a professor at Stanford University School of Medicine, together with his coworker Dr. Chang, performed a clever but scientifically minor experiment that was published in the *Proceedings of the National Academy of Sciences*.[25] The Chang-Cohen paper makes no exaggerated scientific claims. It reports the recombination, inside a living bacterial cell, of DNA fragments from different species. The question in Chang's and Cohen's minds was whether "novel" organisms created by recombining DNA from different species in the laboratory may not in fact arise sometimes in nature, or whether there are species barriers which cannot be crossed without human intervention. This is a serious question that has been repeatedly discussed by Professor R. L. Sinsheimer.[26] Sinsheimer argues that there may be a natural genetic barrier between species which protects the integrity of the species; if these barriers are crossed in carrying out recombinant DNA experiments, he fears that evolutionary processes may be radically altered, and infectious bacteria and viruses may become adapted to infect new species.

The Chang-Cohen paper was widely distributed to Congress before its publication. Since the paper was highly esoteric, it had to be "interpreted" in lay terms for congressmen, and in the interpretation the original results were overextended to imply an unwarranted conclusion. Cohen was quoted in the *Washington Post* as saying "it turns out that Mother Nature has been capable all along of doing in cells what scientists can now do."[27] But in fact the authors had only shown that bacteria can join appropriate fragments of mouse and bacterial DNAs *after* the proper type of fragments have been prepared and inserted into the same bacterial cell. This came as no surprise to the scientific community, since the bacteria under study contain the very enzymes that are extracted and used in the test tube to make recombinant DNA from different species. In this case the bacterial cell was used as the test tube. Chang and Cohen intervened to help nature by isolating certain specific mouse and bacterial DNA molecules and treating them with an enzyme that cuts the molecules in a way that makes recombination possible. Fragments of this type were mixed with bacteria which had been specially treated so that they would absorb the DNA fragments. Laboratory conditions were then adjusted to optimize the results. The Chang-Cohen experiment was therefore as much an engineered event as is DNA recombination in the test tube, and has essentially no bearing on what may occur in nature. It was also implied that there could be no harm in carrying out DNA recombination on a massive scale in the laboratory, once it was established that interspecies recombination could occur in nature. But unless interspecies recombination also takes place on a massive scale in nature, which is clearly not the case, this comforting conclusion is wrong and misleading. Congress was taken in, however.[28]

While Cohen stood by the truth in writing his scientific paper, he was not hampered by it in his political use of the paper. Undoubtedly he had specific motives in mind. The path from truth to power was rather direct in this instance; as Dr. Curtiss said of Dr. Cohen's activities, "It was one of the most imperious, despicable pieces of political science that I know of."[29]

The third event to influence the course of legislation was a meeting held at Falmouth, Massachusetts, on June 21–22, 1977, a Workshop on Studies for Assessment of Potential Risks Associated with Recombinant DNA Experimentation.[30] The workshop was sponsored by the National Institute of Allergy and Infectious Diseases and the Fogarty International Center of the NIH, and chaired by Sherwood Gorbach, professor of Medicine and Microbiology, Tufts University School of Medicine. The approximately fifty invited participants and observers, from the United States and abroad, represented experts in infectious diseases, enteric bacteriology, epidemiology, gastroenterology, endocrinology, immunology, bacterial genetics, and animal virology. Two main topics were discussed during the workshop: (1) the biology of E. coli and (2) the assessment of risks entailed in the use of E. coli as a host for recombinant DNA. In a letter written after the meeting to the director of the National Institutes of Health, Dr. Gorbach stated:

The participants arrived at unanimous agreement that E. coli K12 cannot be converted into an epidemic pathogen by laboratory manipulations with DNA inserts. On the basis of extensive studies already completed, it appears that E. coli K12 does not implant in the intestinal tract of man. There is no evidence that non-transmissible plasmids can be spread from E. coli K12 to other host bacteria within the gut. Finally, extensive studies in the laboratory to induce virulence in E. coli K12 by insertion of known plasmids and chromosomal segments coding for virulence factors, using standard bacterial genetic techniques, have proven unsuccessful in producing a fully pathogenic strain.

Dr. Gorbach stated in this letter that "these remarks are my own impressions reinforced by reading the transcript and the printed materials." In spite of this covering statement, a number of the workshop participants were annoyed at having their views exposed without their prior approval. Their dissatisfaction was understandable since much of the data was fragmentary and preliminary. Their fears were justified, for subsequent experimentation has contradicted some of Gorbach's statements—but not before the letter had been distributed widely, particularly among congressmen and officials of professional societies. In a short time the Gorbach letter had the apparent imprimatur of the scientific community, carrying with it the weight of authentic scientific publication. Senator Adlai Stevenson III had excerpts from the letter put into the *Congressional Record*. He also put into the *Record* the following comments:

> . . . the experimenters actually are saying that these DNA recombinant experiments pose little, if any, risk and therefore there is no need for strict regulation.
> The recent evidence of the decreased risks associated with recombinant DNA research using E. coli K12 as the host vector requires us to weigh carefully the benefits of the proposed regulations against their likely impact on the freedom of scientific inquiry.

There seems to be little doubt, then, that the discussions held at the Falmouth Workshop had an influence on the legislative situation, whatever the participants may have intended.

One of the invited participants later wrote to Dr. Fredrickson, Director of the NIH:

> At this workshop there was, indeed, a consensus that it was unlikely that E. coli K12 could be transformed into a pathogen or would be the cause of runaway epidemics. To be sure this is a very comforting conclusion, but it is only

part of the story. There remains the possibility that the recombinant DNA carried by these "safe" hosts could be
transferred to more invasive strains of E. coli, other enteric
bacterial species or even the somatic cells of their metazoan
hosts. There was considerable discussion about the transfer
of bacterial plasmids and a general feeling that the rate of
infectious transmission in the intestines of healthy mam-
mals would be low. However, in my impression there was
absolutely no consensus reached which suggested that the
probability of transfer of chimeric DNA by plasmids was
sufficiently low to be disregarded.

   I *don't* believe that evidence available through existing efforts at risk assessment is sufficient to justify a relaxation of
the current NIH guidelines on recombinant DNA research
or of the efforts to enforce them.[31]

Two other participants also wrote to Dr. Fredrickson:

Though there was general consensus that the conversion of
E. coli K12 itself to an epidemic strain is unlikely (though
not impossible) on the basis of available data, there was *not*
consensus that transfer to wild strains is unlikely. On the
contrary, the evidence presented indicated that this is a serious concern.[32]

The sequel to these events has justified the cautious participants, for, as Dr. Curtiss noted in his letter of May 11, 1979,
to the NIH:

Since 1977, a number of studies have been conducted which
indicate that the overall probability for transmissions of recombinant DNA from E. coli K12 hosts and vectors is
higher than I or others believed. . . .if the participants at the
Falmouth conference had been aware of these data, more
consideration would have been given to possible consequences of transmission of recombinant DNA to indigenous
[wild] microorganisms of various natural environments.

The Falmouth affair represents still another path from truth to power. The scientific data were fragmentary and uncertain but were made to appear less so; the report seemed to carry the weight of some fifty conference participants, as well as the approval of the scientific community (which had not even seen the data). I personally know that at least five of those who attended the meeting would not have wished their names associated with this maneuver. The Gorbach letter made good newspaper copy; it got into the *Washington Post* and *Science* magazine. But, strangely enough, the transcript of the proceedings was not available to scientific observers until long after the release of information to the press.

The fourth event that altered legislative proceedings for the regulation of recombinant DNA technology was frankly political in nature; actually it was a series of coordinated actions that resulted in an extremely effective lobby.[33] One of the principle organizers was Professor H. O. Halvorson of Brandeis University. He enlisted professional help from the American Civil Liberties Union and organized a coalition of some twenty scientific societies[34] and many prominent individuals.[35] In the Senate, the coalition worked primarily through Senator Gaylord Nelson, who proposed to introduce a new bill in the form of a substitute amendment to the Kennedy bill. The teeth of the Senate bill were thereby extracted.

To gain further congressional support, a massive campaign of telegrams, editorials, letters, and phone calls was initiated during the summer months of 1977. The lobby made free use of the Curtiss and Gorbach letters and of various *ad hoc* arguments. For example, a petition was circulated and signed (under a certain amount of pressure) by most of those attending the Gordon Research Conference on Nucleic Acids of June 1977. It stated:

We, members of the 1977 Gordon Research Conference on Nucleic Acids, are now concerned that legislative measures now under consideration by congressional, State and local authorities will set up additional regulatory machinery so unwieldy and unpredictable as to inhibit severely the further development of this field of research.

. . . the experience of the last 4 years has not given any

indication of actual hazard. Under these circumstances, an
unprecedented introduction of prior restraints on scientific
inquiry seems unwarranted.[36]

This statement overlooked the fact that most of the hazards of
recombinant DNA were expected to be slow in manifesting
themselves, difficult to trace to their cause, and relatively rare
in occurrence. The experience of a few years during the early
development of the technique, before heavy laboratory or industrial use, is therefore negligible. The same argument would
also have justified exposure to asbestos thirty years ago; this is
the kind of thinking that led to the widespread but uninformed
use of estrogens, the contraceptive pill, DDT, and countless
other agents whose side effects did not become known for
some years.

The prestigious, popular, and influential magazine *Science*
also got into the act. The editor stated in an editorial:

The scientists, however, underestimated the publicity
dynamite in DNA. They did not foresee what the media
could do with a topic laden with emotion. They did not
foresee that public alarm could lead toward what some have
called frightening legislation. The clamor reached a peak
earlier this year. During the growth phase, a small band of
scientists were alone in trying to avoid excessive regulation
of their research. What has changed the atmosphere has
been the emergence of a large amount of information about
K12.

During the past few months there has been a remarkable
shift and crystallization of opinion. Suddenly the molecular
biologists have become nearly unanimous in opposing features of the new federal legislation.[37]

To the casual reader it would appear that the "new" information to which the editor alluded must indeed have been
massive. While it is certainly true that new information had
come to light, it was both meager and, as subsequent events
have shown, misleading. Even as it stood, the new data (embodied in the Curtiss and Gorbach letters) did not obviate the

necessity for legislation. A standard scientific approach would have been to consider the data in its entire context, carry out further experiments designed to assess the risks more explicitly and to test safety procedures whenever that could be done, and only then define carefully the limited areas where the results justified a go-ahead signal. Instead, political polarization prevented serious scientific evaluation of the data, and the go-ahead preceded further testing.

The editor of *Science* continued the crusade with further comments on recombinant DNA legislation in an editorial some months later:

> During 1977 the scientific community escaped a threat to the freedom of inquiry in the form of harsh legislation. The ostensible target was alleged hazards of recombinant DNA, but objectives of some of the proponents were broader. The escape from restrictive legislation may prove to be only temporary. Last year congressional action was delayed in part as a result of extremely effective lobbying by scientists, especially a group headed by Harlyn O. Halvorson. If biologists relax the battle could be lost. Moreover, irresponsible acts by individual scientists could be very damaging.[38]

Freedom of inquiry apparently supersedes freedom of conscience, as well as social responsibility, in this narrowly pro-science position taken by *Science,* a very influential publication of the American Society for the Advancement of Science.

In the fall of 1977, Senator Stevenson belatedly entered the combat and announced that his Subcommittee on Science, Technology and Space would hold oversight hearings. As a result of this announcement, as well as the threat of the Nelson amendment and the loss of substantial support in the Senate, Kennedy withdrew S.1217. The only remaining serious contender was the Rogers bill (H.R.7897), which was later blocked at the end of the 95th Congress as a result of Stanley Cohen's activities.

During the lobbying, pressures to escalate recombinant research continued to build within the science establishment. Many American scientists were encouraged to sojourn in European countries where restrictions on recombinant DNA research were virtually nonexistent, especially in Switzerland, France, and Germany; the molecular biologists of the Soviet Union also got into the act. In addition, pharmaceutical firms were quietly but persistently nudging congressmen and enticing molecular biologists with consultantships and lucrative schemes; new firms were formed overnight. In response to pressures from public interest groups, the NIH expanded its Recombinant DNA Advisory Committee (RAC) to include nonscientists. This apparent balancing of forces was actually a power play, for the scientists could and easily did "snow" the newly appointed lay minority. Nevertheless it appeared to the outside world, especially to the lay press, that things were in order and proceeding in a democratic fashion.

As might have been anticipated, the director of NIH and the new RAC set about almost immediately to undercut the Guidelines. The first revision occurred in December 1978, followed by a series of stepwise revisions lowering the safety requirements. Six months later a proposal was made at the RAC meeting of May 1979 to eliminate the Guidelines altogether. As that would have been a politically naive move, the more astute members of RAC opted instead to continue the Guidelines in watered-down form. Eventually, in September 1979, the Guidelines, as an effective tool, bit the dust. Of twenty-five RAC members, only fifteen were present during the momentous vote, and of these, only ten favored the move. The Guidelines now exist in name only. It is an open secret that the demise of the Guidelines was engineered by several influential members of the RAC, in spite of the new experimental evidence (generated by NIH-sponsored programs) showing that several types of risk are considerably greater than had been supposed when the Guidelines were first drawn up. For example it has been shown that bacteria containing recombinant DNA remain alive in humans 500 times longer than had previously been estimated,[39] and that a recombinant containing cancer virus DNA can produce tumors in mice.[40] The British journal *Nature* published a commentary on the serious implications of these experiments,[41] which should have led to an intensification of risk-assessment studies rather than a weakening of the Guidelines.

The lobby of spring-summer 1977 and the subsequent activities of the RAC unquestionably added a new dimension to the path from truth to power in biomedical research; or perhaps it could better be described as the path from "half-truth." Few of those involved can claim innocence in this act. How unfortunate that the first major involvement of the biological research community in a public issue was based not on sober analysis but on a self-serving manipulation of facts! What had started out as a valiant recognition of public hazards on the part of the responsible scientists had been subverted.

In his book on the recombinant DNA controversy, John Lear, editor of the Science and Humanity Supplement of the *Saturday Review,* analyzed the situation as follows:

[The scientists were sure that they] knew what was best for the public. At certain times in response to certain pressures, and to a limited degree, they acted in public view. But they did not at any time seek the public's concurrence in their decisions. They assumed that evidence acceptable to them would bring about concurrence. They were too preoccupied with their own opinions to recognize that their once automatic formula is not acceptable today. They were no longer preoccupied but arrogantly presumptuous when, backlashed by the effects of their original misjudgment, some of them sought to tip the scales by manipulating the evidence.[42]

The original committee established by the National Academy of Sciences in 1974 had set the tone for all subsequent official considerations of the recombinant DNA issue. Paul Berg, of Stanford, one of the first to work with recombinant DNA, chaired the committee and selected its members. It was evident from the composition of the committee that Berg had opted for a narrow base, for only scientists with Establishment views were invited.[43] The president of the National Academy later regretted the absence of "some clinical scientists with long experience in epidemiology and in the handling of genuine pathogens."[44] Conspicuously absent from the deliberation were ethicists, environmentalists, public health experts, and just plain people. The guideline draft produced by the committee was successively watered down, although Berg himself said of the original draft that it was so weak as to be "very likely to draw the charge of self-serving tokenism."[45]

The National Academy of Sciences sponsored not only the Berg Committee but also the subsequent international meeting at Asilomar. In all these activities the Academy addressed only the safety aspects of recombinant DNA research; not until the Academy's public forum on research with recombinant DNA, held in March 1977 after the NIH guidelines had appeared and recombinant DNA research was well under way, were societal issues discussed. But the Academy forum backfired; there were many questions and arguments from the floor indicating the desirability of much more caution and reflection. Evidently the powers at *Science* magazine were apprehensive about possible repercussions after the forum, and it is an open secret that they supressed a report on the forum written by Nicholas Wade, one of their own journalists. The National Academy itself neither publicized nor acted on the proceedings of the forum.

Although the scientific and professional societies to which most scientists belong have a modicum of clout in Washington, the National Academy of Sciences (which does not represent the bulk of the national science effort) is by far the most influential channel for political input. A recent book by Philip Boffey, *The Brain Bank of America,* deals with the structure and function of the Academy.[46] Boffey's critique clearly defines the strengths and weaknesses of this prestigious 115-year-old body. The National Academy occupies a unique position; ostensibly an advisory body functioning with the National Research Council and its committees, in the final analysis it often makes policy by default. When the policy decisions are in scientific or technological areas, they may affect the whole of society. As Boffey points out, the influence of the Academy is strong but often hidden, as in the case of putting a man on the moon. Although the Academy was asked to comment only on certain research and technical questions relating to the moon landing, its final technical OK gave the appearance of approval of the program. The Academy was not asked to comment on the societal consequences of the expenditures necessary for this project. Thus the National Academy, unintentionally and indirectly, had a profound effect on a major policy decision destined to affect a significant segment of the population.

The financial structure of the National Academy tends to favor special interests. The Academy responds to requests for help from federal government agencies such as the army, navy, Department of Defense, and so on; the work is carried out by the Academy with the use of funds supplied by the particular agency. While the Academy is unquestionably interested in issuing an unbiased report, it also wishes to remain viable, and this can be accomplished only by not doing too much damage to the feeding hands; therefore potential criticism is sometimes modulated downward. This modulation is accomplished easily by the appropriate selection of committee members, some of whom may have direct liaison with the sponsoring agency; the fact that the Academy may emend its final report after taking into account the comments of the sponsoring agency also suggests another path from truth to power.

Rather than merely *responding* to questions, it seems to me that the National Academy of Sciences, as well as other scientific bodies, should be concerned with *identifying* the important questions involving science, technology, and society. Where problems exist the Academy should seek real solutions, examining the causes carefully and not limiting itself to the technological fix. But this would require a complete overhaul of Academy organization, starting with the role of the president, its virtual czar, who can by default guide the Academy into various directions with little or no input from the members. Also, the weak and sometimes self-serving committees should be eliminated. With funding arranged to guarantee its independence, the Academy could then provide a much-needed service in bringing science to bear on societal problems.

The initial attempts of the National Academy at examining problems associated with recombinant DNA technology were commendable. This was an unusual instance when the Academy helped initiate the process of analysis in advance, rather than simply responding to an entrenched problem. Yet, when it appeared in their eyes that the scientific establishment might suffer, the course of the Academy was reversed by its influential members. This reversal was disappointing to many of us and was a disservice to the scientific community as a whole. The National Academy of Sciences had the opportunity to take a constructive lead in a situation brimming with societal impact, but it chose not to do so.

The growth of power in science is insidious. Scientists do not embark on their careers with the aim of gaining public influence; quite the contrary. In fact, it has been fashionable for research scientists to look down on any pursuit that smacked of practicality. In general they prefer to practice science as "a monument to humanity's intellectual power and freedom—a modern equivalent of the great cathedrals," in the words of the molecular biologist and Nobel laureate S. E. Luria.[47] It was therefore with a shock of momentary disbelief that I heard the opening speaker at a recent international research conference rejoice that biology, like chemistry and physics before it, was at least about to become an economic force.[48] But what seemed to me a sudden reversal of attitude had already been adopted by most of the audience, scientists engaged in research using recombinant DNA. They recognized, at least subconsciously, the latent power to be gained through this new technique with immense technological potential. It was a matter of scale; what might have been considered prostitution, in a different context, was now a matter of pride.

Insofar as the scientific community has been distinguished by the purity of its motivation, its lack of bias and self-interest, to that same extent it has been free of corrupting power. But today power is thrust upon the scientist by the comprehensive knowledge he has gained, as well as by the vast technological influence of science and the nature of our technological society. The scientist can no longer escape into cathedral-building alone—not on the scale of today's science, resting as it does on public support and public expectation. The scientist has to be more socially responsible than that. If science is to be true to itself, power must be rejected in favor of responsibility. The scientist must have a conscience.

8
Freedom of Inquiry

Freedom of inquiry implies that the inquiry does not impose itself upon others. But science, as it is practiced today, is increasingly intrusive; failure to perceive this has given many scientists a distorted view of the rights of science. Often, when science and technology are closely coupled, freedom of scientific inquiry is even taken to imply freedom of technology. One has to be on guard for this. Technological innovations, whether paid for publicly or privately, are designed for public consumption, and this means that they will have societal effects; because of this technology is not entitled to unrestricted freedom.

Science grew slowly for the two hundred years following the Scientific Revolution. Although science became institutionalized and a partner of industry during this time, it was not until the scope of scientific applications began to broaden, after the turn of this century, that freedom of inquiry became an issue. Chemistry and physics were the major perpetrators of the change. A burst of activity occurred with the outbreak of World War I, when scientists and technologists turned their efforts toward nationalistic goals. Great advances were made in the chemical industry, where explosives, toxic war gases, and other chemicals made their debut. Chemists were lauded for their efforts in what became known as the "chemist's war." Scientific technology experienced another thrust in the 1930s, in the field of physics, with the discovery of atomic fission. The eventual creation of the atom bomb placed American physics at the forefront of science. Freedom of inquiry had paid off, but many physicists lived to regret it. For the first time in the modern era, the impact of science was felt by many to be unmistakably ugly. This realization generated a deep sense of guilt in some scientific circles, and a new awareness of the social implications of science. But the admonitions of these scientists against the use of the weapon were not heeded; the creators of the bomb were part of the decision-making process only insofar as they had placed it at the disposal of others.

In the biological sciences it was the birth of recombinant DNA technology in the early 1970s that suddenly brought the issue of freedom of inquiry to the fore. This technology, giving scientists the ability to impose new details on the old and serviceable architecture of DNA, is pregnant with momentous implications for life. Scientists in the field are aware that the impact of recombinant DNA may well be greater than that of atomic energy.[1] The new technology has already catalyzed a radical change of consciousness for some biologists, although the majority still cannot rid themselves of the ingrained view that the scientific domain is sacred and devoid of social content. The battle lines seem to be drawn. On the one hand we have the scientific elitists, and on the other, those who see science as the servant of more fundamental human values: the reductionist view against the holistic one. This is a fundamental conflict pivoting on the ancient question of man's ultimate purpose.

The argument runs like this: Many molecular biologists tell us that recombinant DNA technology offers practical as well as scientific benefits that are too valuable to forego. The practical benefits—in medicine, agriculture, and industry—are still speculative, although several milestones in the direction of industrial application have already been passed. The immediate scientific benefit is the advancement of knowledge regarding the structure and function of DNA in higher cells. This is unquestionably valuable, this abstract aim, but there lurks within it the possibility of spinoffs that will advance practical aims and perhaps even provide new modes of genetic manipulation. That these possibilities are fully and officially recognized is evident from the level of financial support available for basic research involving recombinant DNA techniques. Overtly, scientists have espoused medical rationales for pushing full-speed ahead in recombinant DNA research. Public statements by prominent proponents of the research tend to overstate the case while still remaining vague, as illustrated by the following quotations:

> Recombinant DNA research is our best hope for understanding diseases like cancer, heart disease and malfunctions of the immune system for which the prospects are poor for prevention solely by public health measures.[2]

> For just as the present-day practice of medicine is impossible without a knowledge of human anatomy and physiology, dealing with disease in the future will require a detailed understanding of the molecular anatomy and physiology of the human genome.[3]

But there are also dangers involved in recombinant DNA technology, both grave uncertainties and recognizable risks. Those eager to exploit the new method present its scientific promise as far outweighing the risks, which they see in terms of accidental (or malevolent) release of potentially harmful microorganisms. Having estimated these risks to be small, they consider any questioning of their position as a challenge to scientific freedom and the ultimate value of knowledge.

Others of us feel that, even setting these accidental risks aside, a technique whose implications reach so far beyond the domain of science must be thoroughly evaluated in the entire human context; scientific inquiry is not the only freedom in question. There are those who consider the natural gene pool of the earth to be an inalienable birthright; they are not willing to accept, for themselves or their progeny, a genetic world patterned on other mens concepts of the ideal. This is a political issue, not a scientifc one, and it relates to fundamental freedoms that have never had to be incorporated into bills of rights because man has never before had power over them. Now these rights to the heritage of the earth are threatened by science-based technology, which has outstripped public awareness and political wisdom. No long experience of this kind of slavery has yet sensitized humanity to the threat.

Because of the implicit conflict of rights, other, calmer voices need to be heard above those of the interested scientists. In the words of Harold Green, a professor of law at George Washington University:

> If we desire to guide the development of technology so that we may enjoy its benefits free of unacceptable injury, we must find ways to intervene before momentum takes over. The enthusiasm and optimism of the proponents of the technology must be tempered at an early stage by a more deliberate, explicit, and somewhat more pessimistic consideration of the area of uncertainty as to potential hazards.[4]

There is no doubt that molecular biologists have discovered an incisive tool in recombinant DNA technology. Deeply excited by the scientific possibilities of the technique, they are rushing to put new genes into bacteria, but they are paying little heed to what lies ahead. This precipitancy is not limited to the United States. In November 1977, the federal Interagency Committee on Recombinant DNA Research reported that 300 laboratories in this country, 150 in Europe, and 20 to 25 in Canada, Japan, Australia, and the Soviet Union were doing recombinant DNA research. In 1979 there were more than 700 projects sponsored by the NIH alone. I cannot help but feel an abnegation of judgment here, a submission to the attractions of mere technical virtuosity. The ostensible purpose of this research is to improve the human condition, yet no analysis has been made of the ultimate impact of genetic manipulation on the human condition over the long run. The right to define the meaning of "improvement" in this controversial area is simply being appropriated as a concomitant of power and applied in the interest of short-range objectives.

The scientist's myopic vision in this situation requires a counterforce. For him, the relative importance of science is a subjective feeling; science and the scientific way of thinking constitute a large part of his being. The usual training of a scientist subliminally reinforces certain exalted attitudes about the role of science, its relationship to reality, its special competence to determine what *is* and therefore perhaps also what *should be*. In this way, freedom of inquiry is easily stretched to encompass actions that begin to encroach on the rights of others.

Basic academic research is not intrinsically exempt from this criticism. A scientist who pursues a line of research with obvious technological potential knows full well that in our worldwide society there is no mechanism for reviewing the technology from the standpoint of the common good before it is applied. No one is waiting to accept the social burden from him. The scientist can therefore not escape a large measure of responsibility for the uses made of his discoveries. As always, responsibility defines the valid limits to freedom, including freedom of scientific inquiry. It therefore behooves the academic scientist to think beyond his immediate interests, lest he fuel a fire that neither he nor anyone else will be able to control. I am not speaking here merely of the more obvious form of responsibility where the scientific experiments themselves may be hazardous to the public.

Few academic scientists have paid much attention to the technological significance of their work, except insofar as it may be useful for funding purposes. Many of my colleagues were horrified by the use of chemical and biological warfare— results of scientific research—in Vietnam. Yet, in their own fields, these same scientists feel justified in demanding complete freedom to pursue whatever puzzle occupies their thoughts, limited only by the universally accepted methodologic ethic.

But, the fact is that in many fields of science, and increasingly so in biology, the distinction between technology and science—and therefore the "right" to scientific freedom—is not always clear-cut. For example, the recent manufacture of insulin through recombinant DNA technology by Walter Gilbert and his co-workers at Harvard, and by William Rutter and Howard Goodman at the University of California,[5] involved procedures which had already been established; no new basic scientific information came from this work. It was a sophisticated engineering feat, with results readily transferable to industrial processing. Several companies were ready and waiting for the data. Although the work was done under the guise of basic science at two leading universities, it is certainly not devoid of social content.

With recombinant DNA the practice of technology as science is already becoming blatant, as universities take out patents for industrial processes developed in their scientific laboratories and scientists set up companies[6] to exploit the results of research carried out in academic laboratories with government support. One can question whether this is the disinterested pursuit of pure knowledge, and whether it is entitled to a guarantee of academic or scientific freedom.

In chapter 5 I tried to show how much freedom of inquiry has already been sacrificed as scientific research has come to be conducted on a large scale with a big and powerful establishment, sophisticated instrumentation, and heavy dependence on public (i.e., government) support. This loss of freedom is not the result of conscious choice, but it is generally accepted by scientists, somewhat ruefully, as a fact of life or is even overlooked by those not given to analysis of their situation. I have heard grumbling, but no outcry of protest. If scientists are able to rationalize limitations on freedom of inquiry imposed by a scientific system that is in many respects arbitrary, it is hard to justify a show of outrage when it is a question of the public interest.

Many scientists maintain that no efforts to limit freedom of inquiry can be in the public interest; it is implicitly on these grounds, that freedom of inquiry is claimed as a "right." Hidden in this assertion is the antiintellectual notion that a rule of thumb concerning the relationship of science to the public can take the place of continuing intellectual analysis in the light of new developments. This is an untenable assumption for a profession committed to experimental verification. There is no doubt that determination of the public interest is often a difficult matter, but scientists have not been given the right to sidestep the issue or arrogate the responsibility to themselves alone; if they believe themselves to be uniquely qualified for the task, they had better request it of the public.

Research as it is usually practiced today is an integral part of a state-financed operation aimed at application. Under these particular conditions, one can ask whether freedom of inquiry in all directions is sensible or proper. For biomedical research, public money represents the major support, and many research results will inevitably have a societal impact. The public has a right to a meaningful input in decisions that will affect it. In addition, the unmistakably operational character, however vague, of public programs confers a social responsibility on all who participate in them. Many scientists involved in recombinant DNA research have even more overtly forfeited a valid claim to rights associated with value-free research because they have suggested, and are encouraging, applications of the technology for various purposes, such as hormone manufacture. But even those who are not directly concerned with applications are not free of responsibility; we are not talking of Bacon's vision of three centuries ago, when he foresaw "making new species; drawing new food out of subjects not in use; engines of war stronger than canons," etc. These speculative predictions were to be realized only much later. A scientist today knows *beforehand* that his ideas or processes will be exploited. His entire research effort should therefore be conceived from an enlightened and broadened viewpoint that includes societal values. André Cournand, professor of medicine, emeritus, at the Columbia University College of Physicians and Surgeons, has summarized this view succinctly:

The scientist's code should explicitly take cognizance of the fact that the scientist is an individual who lives in a society which has ends other than the cognitive ends of scientists, and that the cognitive achievements of scientists do not always and necessarily serve these ends. Scientists themselves have multiple allegiances, both within the scientific community and outside it. They need norms to help them find the right balance among these alliegiances.[7]

In the absence of an established, efficacious mechanism to ensure the application of scientific knowledge in the public interest, the search for knowledge itself cannot be considered neutral. The scientist who insists on placing the responsibility with the technologists is rather like a person who makes a useful box of matches and leaves it in a room full of pyromaniacs. After all, it is the scientific community that can best foresee the technical implications of its discoveries and recognize factors that could make the results of application unpredictable or dangerous. Experience has certainly shown that industry cannot be trusted to take into account even all the *known* factors, or to inquire about *obvious* risks.

Yet scientists often protest that it is not their job to evaluate; they prefer to participate in the traditional optimistic expectation of unpredictable future benefits, and to dissociate themselves from the accumulating evidence of technological disaster. The official view maintains that it is antiintellectual to limit freedom of inquiry. In discussing this question in *The State of Academic Science*, Smith and Karlesky observe:

The question of both lay and expert control over research will present issues of increasing complexity. Conflicts of basic values will often be involved. Assessments of the potential harm done by pursuit of certain lines of research are extraordinarily difficult. Fear of the unknown and fear of the consequences of knowing could open the way to at times irrational and even demagogic efforts to control research. On the other hand, the insensitivity of some scientists in their insistence on the primacy of a single set of values has complicated efforts to arrive at reasonable solutions.[8]

In recent times it has become fashionable in scientific circles to take the position that *all* knowledge should be pursued; one might call it the Mt. Everest syndrome. This syndrome was excusable for Bacon or Descartes, who perceived no limits to the manipulation of nature by man; excusable because the means of implementation were not at hand. Neither of these men, nor their philosophies, threatened society in their time. But in our technological society, with its substantial scientific resources, such an attitude smacks of arrogant foolishness. As I discussed more fully in previous chapters, the limits to man's power over nature and the consequences of pushing toward those limits are now visible and looming over us. The scientist who does not perceive this today must be looking the other way. Science can be employed to advance or push back the cataclysm: are both paths equally to be encouraged?

In my view there is no question that science, as well as technology, requires a certain amount of societally oriented guidance at this point in history. It is not this principle but its implementation that should concern us. Harold Green has commented:

The problems under discussion at this Asilomar Conference are not unique. Comparable problems of balancing benefits against risks are found in many other areas in which science and technology are advancing. One element that is common to all of these areas is the fact that benefits are always relatively obvious, immediate, and intensely desired, while risks are usually relatively remote and speculative. There is, moreover, usually no constituency for the risks—very few people have the knowledge, resources, or incentive to press the risks upon decision makers. Our major need is to find means through which risks are given time and dignity more equal to that given to benefits in the decision-making process.[9]

In a similar vein, Harvey Brooks of Harvard, chairman of the congressional panel on the Health of the Scientific and Technological Enterprise, has said:

Until recently, society has acted on the principle that new technology should be assumed innocent until proven guilty. This was reasonable when technology was less powerful than it is today, but with time the price has crept up and the burden of proof has shifted much more onto the advocates of the introduction or diffusion of technology. Lack of evidence of side effects may now be sufficient reason for deferring a project whose possible secondary effects are not fully known or understood.[10]

A risk-benefit analysis implies that the technology is already upon us and that we somehow cannot avoid it. It would be better to anticipate this dilemma and perhaps eliminate the problem at an earlier stage by not developing certain questionable techniques. Rather than viewing the situation as one in which freedom of inquiry has to be stifled, it would be more constructive to seek actively the application of scientific intelligence in the directions most important for society and in ways that offer the best hope of producing useful results with a minimum of attendant dangers. This means a reevaluation of needs, which might entail the sacrifice of some freedom of inquiry: a tradeoff for more basic values—perhaps for survival.

The enlightened and informed consent and participation of scientists is needed to set priorities for a new *modus vivendi*. To start with, the scientist should feel morally bound to see to it that there is an intelligent direction to his curiosities; he should no longer simply offer his wares to the technologists. He should choose approaches which do not have the potential for irreversible damage. Recombinant DNA technology is only one pathway to the solution of certain biological problems; there are others. We set priorities all the time in the rest of society. Should not scientists be bound by the same rules? In a recent article, Harold Green noted that there are already a variety of valid legal restrictions on research, such as those regarding experimentation on human subjects, the use of certain drugs, and so forth. He commented: "It is not clear to me why, in the face of these precedents, the scientific community has become so edgy about scientific freedom re recombinant DNA in recent months."[11]

The edginess is illustrated by a resolution entitled "An Affirmation of Freedom of Inquiry and Expression," passed by members of the National Academy of Sciences on April 27, 1976:

I hereby affirm my dedication to the following principles:

That the search for knowledge and understanding of the

physical universe and of the living things that inhabit it should be conducted under conditions of intellectual freedom, without religious, political or ideological restriction.

That all discoveries and ideas should be disseminated and may be challenged without such restriction.

That freedom of inquiry and dissemination of ideas require that those so engaged be free to search where their inquiry leads. . . .

The resolution omits any mention of responsibility, and makes the implicit assumption of the neutrality of science. It does not take up the question of financial patronage, which has a lot to do with the direction of inquiry. It is therefore a document without much relevance to the real world. The unqualified freedom of inquiry longed for by these scientists is not a "right" but a vestigial privilege created by societal decisions in the past. No first principle declares this "right" to exist under all circumstances; there are many legitimate limitations to freedom. In this light, as Professor Green puts it: "While an argument about a right to scientific freedom may be a useful piece of rhetoric in a political debate, we should not take the existence of such a right too seriously."[12] Kurt Mislow, a professor of chemistry at Princeton and a member of the National Academy of Sciences, has courageously cut through the prevailing myths to get at the heart of the matter:

I will undoubtedly provoke cries of inquisition and the like, but I must nevertheless force myself to say that I don't agree that freedom of inquiry should be limited only if actual hazards are preceived. I do not agree that increased human knowledge is of paramount importance. I do not agree that the real enemy is ignorance. I think these are trademark shibboleths which everybody accepts without qestioning. I can think of lots of examples where knowledge is extremely dangerous. And in the search for knowledge, you have to ask what you are going to do with the knowledge once you have acquired it.[13]

The too often quoted quisition of Galileo is not relevant to the present situation. Galileo was censured for his use of the scientific *method* itself. This is not in question today, with recombinant DNA technology or any other research endeavor. Accusations of Lysenkoism are equally beside the point, for in that case, the Soviet government attempted to legislate truth.[14] Philip Handler, president of the National Academy of Sciences, also misstates the case when he says, "The objective of some who have proposed regulation for recombinant DNA research is to use the power of government for the suppression of ideas that may otherwise flow from such research."[15] On the contrary, all the proposed regulations have been based on the guidelines drawn up by the National Institutes of Health and on similar proposals for the regulation of research with respect to safety in the laboratory, in the community, and in the environment. No one has even implicitly supported the regulation of ideas, although I, and others, have advocated the regulation of technological applications in the public interest. Statements such as Dr. Handler's only serve to muddy the waters. I believe that we have reached a point in history when the highest and most socially useful exercise of human intelligence should be shifting from the search for new information to the wise evaluation and control of what is already known. This can and should be encouraged by government; it is not censorship but social responsibility.

People imagine that the scientist deals only in facts whose validity can be demonstrated; his statements to the general public therefore tend to take on the character of infallible pronouncements. His personal and human frailities are neatly obscured beneath this professional facade, and the extent to which judgment and predilection play a role in his thinking is not generally appreciated. People have been taken in; they have acquiesced in giving scientists, both basic and applied, full rein—freedom of inquiry and also of technology. Bentley Glass, a dean of contemporary science, has put it poignantly:

Indeed, so awesome is already the accelerating rate of our scientific and technological advance that simple extrapolation of the exponential curves shows unmistakably that we have at most a generation or two before progress must cease, whether because the world's population becomes insufferably dense, or because we exhaust the possible sources of physical energy or deplete some irreplacable resource, or because, most likely of all, we pollute our environment to toxic, irremediable limits. Many scientists have in recent decades examined these processes and tried to flag the runaway express. Let me suggest at the risk of grave misunderstanding that in future histories of the world the decade of the 1960's may be known not significantly for the miserable Vietnam war but as the time when man, with unbridled lust for power over nature and for a so-called high standard of living measured by the consumption of the products of an industrial civilization, set in motion the final speedy, inexorable rush toward the end of progress.[16]

The benefits we have reaped through science have been many. Anyone who overlooks them or plays them down is overly cynical. But we simply have to stop demanding every toy in sight, like so many spoiled children. The trouble is that there are no built-in control mechanisms in the technological system; the Western worship of free enterprise and the socialist emphasis on industrialization have taught us to spurn them. However encouraging the initial results of laissez-faire science and technology may have been, it is now clear that society has turned the corner toward unabated technological decadence. It is not in our own best interests to allow total freedom to all members of the scientific community. This does not mean that all science should feel the oppressive weight of public control; that would be counterproductive. But it is time to take a second look. I think that the kind of freedom of inquiry expected by scientists today represents a diffuse and vague longing that harks back to earlier times; it is encouraged by the persistence of the science establishment in conveying the false notion that scientists can have it both ways: funds on a large scale and freedom. Scientists would be better off if they faced this issue squarely and encouraged and participated in the creation of frankly targeted, well-defined programs along with a modicum of untargeted ones. Pure research has seldom been carried out on a large scale, yet it has sufficed to provide us with many applications. A continuing small-scale effort ought surely to be encouraged as a cultural activity, in which freedom of inquiry would be exercised with a conscience.

In the mid-1940s we witnessed the fruit of nuclear research when the first atomic bomb was exploded; and at about the same time we learned that DNA is the genetic substance. Within a matter of months we were presented with two great secrets of nature: one at the core of matter, the other at the core of life. These discoveries, qualitatively different from any we had known previously, have greatly accelerated our approach toward new boundaries in human affairs. The sequel will require a lot of soul-searching and the relinquishing of many elements we now think essential. The sacrifices will be easier to make if we recognize that the future of the entire human race is at stake—not only its survival but also its nature and its freedom. We are no longer talking about science and its freedom of inquiry but about human life and liberty and our social responsibility toward them. The imposition of fundamental and irreversible changes on future generations or their environment is unjust, as it is unjust for any particular group to impose its will on others without their consent. In *The Abolition of Man*, C. S. Lewis has put this most forcefully: ". . . what we call Man's power over Nature turns out to be a power exercised by some men over other men with Nature as its instrument."[17] In the last analysis, unrestricted freedom of scientific inquiry is the means of supplying the few technocrats with new modes of power over all the rest of us. Scientists should examine this cryptic consequence of scientific orthodoxy before they knwittingly commit themselves to a socially irresponsible position.

9
Conscience in Science

No matter how you choose to evaluate it, the social price of science-technology today is often exorbitant, perhaps even prohibitive in the last analysis. Yet, somehow, we hate to admit it. We still nurse the vague hope that the problems will solve themselves without our having to act on them. The amenities that surround us are undeniable, but how many of them are trivial substitutes for the things of value they displace? Can we sort the good from the bad? Are we in fact really free to select and enjoy only what we consider desirable?

When this or that technology begins to usurp our freedom the transgression is likely to be subtle and sometimes unintended, yet we feel its existence and certainly we know it when the job is done. We are beginning to recognize, for example, that the subjugation of the environment, with its potentially devasting effects on humankind, is a major catastrophe of our time—an incalculable loss of not one but many freedoms: the freedom to breathe clean air, or to drink pure water. Little solace is to be found in the knowledge that man has been changing the environment by agriculture and domestication of animals for millennia. What's more, the price of continuing technological innovation includes a series of inevitable fringe benefits such as the fragmentation of relationships at all levels of society: between individuals, within the family and the community, between industrialized and underdeveloped na-

142

tions, between man and nature. Somehow, in recent times, we have organized things so that accelerated, destructive change has become necessary for economic survival. We are urged, sometimes forced, to use all manner of automation in the name of economic efficiency, but at the expense of one of our greatest needs, human involvement. Technology intervenes even in the simplest functions of daily life, robbing us of the human component, usurping our freedom of choice. Not only are we losing control over our individual activities but, worse still, over the computerized giant that searches out the smallest niches in our lives and operates inside them.

The oppressive weight of technology together with the capability we have developed for ultimate destruction by nuclear war make an awesome combination that has created a feeling of impotence and deep frustration in almost all of us and a sense of hopelessness in many. A long list of social philosophers, cultural historians, and other scholars have observed and analyzed the situation in depth.[1]

The economist Robert Heilbroner summed it up concisely and most poignantly in the opening sentence of his book *An Inquiry into the Human Prospect*: "There is a question in the air, more sensed than seen, like the invisible approach of a storm, a question that I would hesitate to ask aloud did I not believe it existed unvoiced in the minds of many: 'Is there hope for man?' "[2]

An affirmative answer surely calls for a superhuman event. An optimist might imagine that this would take the form of a sudden awakening to social consciousness by a critical mass of humanity—a mass so large and so single mindedly determined to change things as to cause the technological system to give way and collapse. Yet, this possibility is almost precluded by the inner logic of the system, whose main principle is self-preservation and expansion of its aconscious state. But I believe that scientists, with their tradition of independent intellectual analysis, will be among the first to awaken, and perhaps they will create an effective nucleus that will crystalize the awareness of all. In this chapter I discuss the social overtones, both encouraging and discouraging, of some of the present views and attitudes of scientists.

There is a contemporary fashion for touting scientific achievements as great benefits for mankind (even though public relations efforts in science have often backfired). Recombinant DNA has been widely promoted in this way. John Lear has commented:

> The measurable long-range value of gene splicing is necessarily debatable. But the immediate rewards for the involved scientists, in terms of their incomes and career advancement, are not. It is only natural for them to put their most optimistic foot forward (although, if they expect to maintain credibility in the public mind, they should present their claims in the normally accepted ways of science), but the rest of us would be irresponsibly gullible if we did not rigorously examine the reasonableness of their requests for our support.[3]

While many of the possible benefits of this technology seem attractive at first glance, experience warns us of possible deception. But calls for a cautious approach have been drowned out by accusations of antiscience and demands for freedom of inquiry. Scientists have rushed ahead at full speed, and three preindustrial test products of recombinant DNA technology are already before us: insulin, somatostatin and interferon. Choices regarding the application of technology are being made *de facto*, and neither you nor I have any meaningful input in the process. The scientists' impatience is preempting public choice.

Two possible applications of recombinant DNA that have attracted wide attention are in agriculture and gene therapy. Both claims, as scientists admit, are premature from a scientific point of view, yet as in the case of the war on cancer a few years back, word has gotten around and feelings have been generated that, somehow, we can expect panaceas in these areas. In agriculture, recombinant DNA technology is to be used to transfer nitrogen-fixing genes from bacteria to plants so that atmospheric nitrogen can be directly used by food crops, thereby obviating the need for nitrogenous chemical fertilizers. This would increase food production while decreasing its energy-dependent cost. It might also create serious ecological problems by affecting the chemical composition of soil and waters, or changing the interdependent relationships between life forms. These are the hazards of forcing evolution to take quantum jumps instead of the tiny steps that occur in nature. An enormous amount of research will be required before the goal can be realized, if in fact it is achievable. One reason for this is that nitrogen-fixation genes can only function in cells with an exceptionally high energy output, and so it may be necessary to alter the entire metabolism of the plant, thereby radically changing the character of the food crop. But assuming success in the venture, one can still ask whether altering plant metabolism is the solution to the problem of hunger and starvation in the world. I want to outline some details of this problem to show how far off the mark the proposed technological solution is and, consequently, why the new technology is in this instance irrelevant.

The world population is now about 4 billion. It is expected to increase to about 8 billion near the turn of the century, thence to about 16 million in 2040.⁴ Although we have the theoretical capacity to feed all the 4 billion people alive today, nevertheless, in the underdeveloped countries a billion people receive too few calories (according to a World Bank report). If we were to use less grain for cattle feed, there would be adequate food for everyone. Even under these circumstances, however, many people would remain undernourished because of growth and distribution patterns that depend on political and economic factors.

One of the fundamental limits to food production is of course the amount of available arable land, which is estimated at approximately 8 billion acres. Each person requires about one acre (including space for food production, roads, industry, living space, and disposal of pollutants) in order to live at a civilized level. The comfortable limit on world population size is thus automatically set at approximately 8 billion, *if everyone eats a cereal diet and resources are equitably distributed.*⁵ These figures are generally agreed upon by ecologists and food experts. In any event, even if the estimates are off by a factor of 2 (i.e., if the earth can support 16 billion people), we are already perilously near an uncomfortable limit. Now is the time to act, for it will be too late when the events are upon us in about fifty years. If we wait until the population level becomes critical, vast numbers of people will die or be killed or undergo terrible hardships until the population density is decreased.[6]

What has to be done is to stabilize the population before it reaches a calamitous level and, in the meantime, to reorganize the utilization of our present agricultural resources to eliminate the waste now prevalent and the great inequities in food distribution. Beyond that, there is a limit to how far the catastrophe can be pushed back by increasing productivity. In the United States we are already close to the theoretical limit of what an acre of land can be made to produce.[7] Throughout the world, factors such as water supply impose a natural limit; the UN Food and Agricultural Organization has forecast a serious water shortage by the year 2000. Moreover, agriculture is in ecological competition for natural resources used for other purposes: water is needed for power plants, factories, mines, and other types of production; land is continuously taken for highway and other nonagricultural uses. In addition, climate is an uncontrollable variable; in bad years, crop yields can drop perilously low.

A systematic analysis of the important variables—resources, food production, industrial production, population growth and pollution—has been carried out under the aegis of the Club of Rome. The results have appeared in the book *The Limits to Growth,* wherein the authors show that it is possible to achieve a near-equilibrium state in which the rights and values of both society and its support systems are respected; however, certain current values would have to be altered, in order, for example, to produce a shift in emphasis from material products to services such as education, health and so on. The suggested approach yields a steady state for the last four variables noted above; the first—resources—projects only a small decline into the twenty-first century. This would allow for a transition period during which new, appropriate technologies could be developed. *The Limits to Growth* emphasizes the runaway nature of exponential growth, the kind we are experiencing in developed countries. Analysis of the world system as a whole shows that we are deceived in believing that there are hundreds of years left in which to make the necessary adjustments to impending resource depletion. For example, the authors show that appropriate changes begun in 1975 could lead to an equilibrium state with no great mishaps, but that if the same changes were instituted in the year 2000, an irreversible and calamitous decline in all support systems would be unavoidable.[8]

Although the confirmed technologists will opt for business as usual and hope for technological panaceas, these views are unrealistic if only because there is not enough time to develop the sweeping new technologies that would be required. And no amount of scientific and technological sophistication would buy more than a small amount of time—at the end of which, the situation is likely to be worse, as the population continues to grow. If the threat is not great enough to produce action now, no amount of time will bring the species to its senses until disaster hits.

It is obvious that it would be naive to attempt to solve the food problem with recombinant DNA technology. Even if that technology should someday succeed in producing plants that can fix atmospheric nitrogen, the most that could be hoped for would be a small contribution to a temporary respite—a technological fix that has no bearing on the fundamental population problem and might have adverse side effects that would exacerbate the situation by producing ecological instability. The real solution can only be one based on a true equilibrium between utilization and production, not an ever-increasing exploitation of the limited resources of the earth.

In my view scientists ought not to participate in the development of technological fixes of questionable value, especially when their efforts and resources are so needed elsewhere. As former Secretary of Health, Education, and Welfare Joseph Califano emphasized:

. . . we must make choices. It is a hard fact but a reality that not every area of basic research, perhaps not even every promising one, can be explored at once or with equal energy and equal commitment of resources.[9]

A more urgent direction for research related to agriculture is in the application of scientific principles to ecological farming. Such farming relies on solar energy, more human input, recycling of wastes, and intensive land use. Its aim is maximum production with minimum energy consumption. This new way of farming would ensure the integrity of the earth's ecosystems for the indefinite future. The risks-benefit tradeoff is overwhelmingly in favor of this type of research, compared with the genetic engineering of plants through recombinant DNA.

The application of recombinant DNA techniques to human gene therapy has also been widely discussed. At a symposium in 1977 a Nobel laureate outlined a detailed hypothetical protocol for a specific type of gene therapy. He prefaced his description thus:

> There are still many people who do not believe that genetic engineering is feasible, so let me offer to you a possible scheme to indicate how close we could be to attempts at genetic engineering.

He then described an experimental protocol in which the desired genes could be introduced into bone marrow cells outside the body; these cells can then be re-introduced into the body of an afflicted individual. Then came a note of uncertainty:

> I have little doubt that within five to ten years just such a experiment will be attempted, and that, if it is successful, gene therapy could be added to the arsenal of hematologists. I should point out that this scenario assumes that the newly added gene will function normally and under appropriate regulation in the cell which receives it. There is certainly no guarantee of this, but it seems likely that animal experimentation could teach us how to provide the genes in an appropriate manner.

Then he threw in:

> I should also point out that this will generate something of

a new industry if it comes about, because there will have to be people who know how to make the genes, to link them, and to provide them in therapeutically useful form.[10]

In spite of the speaker's soft pedal, the net effect of this presentation was to indicate great promise of progress toward a laudable goal. The serious flaw in this exercise is that the scientist describing the science did not give equal time and emphasis to its social and ethical aspects; the science therefore rolls along in its "value-free" way, in this case toward a very specific aim of application. This is not "pure research," aimed only at knowledge.

After having wooed the public's approval and support with this carrot on a stick, can that scientist really claim, after the techniques have been developed, that any ill effects that may arise from their utilization are not his responsibility?

First, let us consider whether this is really a high-priority line of research that deserves to be chosen in preference to other directions. There are some 2000 known genetic diseases, all relatively rare among the human population. A completely different type of gene therapy would have to be developed for each disease, since in each case replacement genes would have to be inserted into different kinds of cells. The cost of development and, more important, the cost of applying gene therapy would be high, so high that the therapy would probably not be available to most sufferers from the diseases. A large fraction of genetic diseases would have to be treated at the early embryo stage in order to be successful; but if the defect could be identified at that stage, the logical treatment would be abortion. And successful treatment of postnatal individuals by gene therapy would not improve the human gene pool because the defective gene, rather than the inserted one, would be passed on to succeeding generations. In principle, however, gene therapy could be carried out on germ cells, a procedure that would permit the transmission of inserted genes to offspring. This is the fond hope of molecular eugenicists. At this point we leave behind the concept of "therapy," for the same techniques can be used to insert any kind of gene, not just replacements for defects.

So, with the image of aid for child victims of genetic disease in our minds, we find ourselves on the doorstep of eugenics. The aim of eugenics is to increase the frequency of "desirable" characteristics in human progeny. Eugenics was in vogue about fifty years ago, when it received considerable support from industrialists like the Harrimans, Kelloggs, and Carnegies. Fortunately, the peak activity of this movement was relatively short; its demise was aided by the repugnancy of emerging Nazism. The scientific establishment did not take a strong position on the issue, and so it has remained dormant. With the development of new genetic techniques, the eugenics movement in America could emerge again. Jonathan Beckwith of Harvard Medical School has warned that several eugenic techniques are already in use.[11] The techniques include amniocentesis and postnatal and adult genetic screening. These procedures undoubtedly benefit a few individuals, but they are clearly open to abuse. The classical question arises: who decides what is a defect? Today the burden is on practitioners and recipients of the techniques; but the day is approaching when social and political forces will play a more significant role in "private" decisions. But there is a more insidious aspect to be considered. The use of these apparently beneficial genetic procedures creates an atmosphere in which genetic procedures in general become an accepted form of solution for many problems. But, as Professor Beckwith points out, many of these problems have strong social and political components; genetic techniques become a rationale for tolerating the situations that generate the problems. Beckwith writes:

In the United States over the last few years, approximately 1 million school children per year have been given drugs, usually amphetamines, by the school systems, in order to curb what is deemed disruptive behavior in the classroom. It is claimed that these children are all suffering from a medical syndrome, minimal brain dysfunction, which has no basis in fact—no organic correlate. Now clearly, there are some cases of children with organic problems where this treatment may well be important. But in the overwhelming majority of cases the problems are a reflection of the current state of our crowded schools, overburdened teachers and families, and other social problems rather than something wrong with the kids. Imagine, as biochemical psychiatry is providing more and more information on the biochemical basis of mental states, the construction of a gene that will help to produce a substance in human cells which will change the mental state of individuals. Then, instead of feeding the kids a drug every day, we just do some genetic surgery and it's over.

Another more recent example of this genetic approach to social problems lies in the field of industrial susceptibility screening. Arguments have been appearing in the scientific literature and elsewhere that occupational diseases, caused by pollutants in the workplace, can be ascribed not to the pollutants themselves, but to the fact that some individuals are genetically more susceptible to the pollutants than other individuals. So the argument goes, the solution is not getting rid of the pollutants but rather, for example, simply not hiring those individuals who are thought to carry the genetic susceptibility. . . . A Dow Chemical plant in Texas has instituted a large-scale genetic screening program of its workers. Rather than cleaning up the lead oxide in General Motors plants, women of child-bearing age are required to be sterilized if they wish employment. It is a genetic cop-out to allow industries to blame the disease on the genetically different individual rather than on their massive pollution of the workplace and the atmosphere. This is the epitome of "blaming the victim."[12]

Genetic screening in petrochemical plants is on the increase and threatens to become a standard procedure for detecting "defective" genes in workers who are thereby labelled "hyper-susceptible." OSHA and the Equal Employment Opportunity Commission claim that the Allied Chemical Corporation, B. F. Goodrich, Avtex Fibers and American Cyanamid also exclude women of child-bearing age from certain jobs.[13]

The idea of breeding workmen with genetic resistance to industrial pollutants would probably be laughed off as science fiction, but the much-hailed recent achievement of human *in vitro* fertilization, which produced the "test-tube baby" in Britain, could lead straight in that direction. Experiments attempting to produce hybrid mice, using recombinant DNA before reimplantation, are already underway. Human cloning is already being regarded with curiosity and interest. A human embryo in the test tube lends itself admirably to cloning and genetic engineering procedures (including those proposed for gene therapy). Who knows what new and useful human characteristics could be developed by research in this area? Or dare we hope that scientists will conclude that the genetic rights of future generations take precedence over scientific freedom and the passing demands of technology?

Eugenics was one of the subjects discussed in 1963 at a Ciba Foundation symposium on the impact of science on the fate of mankind.[14] I give here a brief account of some of the more interesting views because they provide an insight into the kind of thinking that goes on in the rarefied atmosphere of a conference at which great and creative scientists turn their minds lightly to societal issues.[15] Listen.

According to Julian Huxley, "our present civilization is becoming dysgenic" because of modern technology, especially in medicine, so that more humans with genetic defects reach maturity and are permitted to procreate, thereby increasing the genetic load (as it is called). He proposed that we reverse the apparent trend by developing new human reproduction techniques to produce eugenic improvement of the species. J. B. S. Haldane, the eminent chemist-geneticist, suggested cloning people; he would use donors of proven accomplishments who were at least fifty years old (except for dancers and athletes, who would be cloned younger). Embryos would be grown in a culture, in as many copies as seemed desirable. If the clonal progeny of Arthur Rimbaud were to grow up with no propensity for poetry, and became second-rate empire builders instead, the clone would not be grown further. People with rare capacities, for example, permanent dark adaptation, lack of the pain sense, and special capacities for visceral perception and control, as well as centenarians (if reasonably healthy), would generally be cloned, not that longevity is necessarily desirable, but data on its desirability are needed.

Haldane was not reading from *Brave New World*. Undoubtedly he was inspired by Hermann J. Muller, the Nobel Prize winning geneticist who had earlier suggested eugenic approaches in his book *Out of the Night*. At the Ciba symposium Muller said: "Man as a whole must rise to become worthy of his own achievements. Unless the average man can understand and appreciate the world that scientists have discovered, unless he can learn to comprehend the techniques he now uses, and their remote and larger effects, unless he can enter into the thrill of being a conscious participant in the great human enterprise and find genuine fulfillment in playing a constructive part in it, he will fall into the position of an ever less important cog in a vast machine. In this situation, his own powers of determining his fate and his very will to do so will dwindle, and the minority who rule over him will eventually find ways of doing without him."[16] The last part of this statement brings to mind the Germany of the 1930s.

More recently, the molecular biologist James Bonner has stated:

The logical outcome of activities in modifying the genetic make-up of man is to reach the stage where couples will want their children to have the best possible genes. Sexual procreation will be virtually ended. One suggestion has been to remove genetic material from each individual immediately after birth and then promptly sterilize that individual. During the individual's lifetime, record would be kept of accomplishments and characteristics. After the individual's death, a committee decides if the accomplishments are worthy of procreation into other individuals. If so, genetic material would be removed from the depository and stimulated to clone a new individual. If the committee decides the genetic material is unworthy of procreation it is destroyed. . . . The question is indeed not a moral one but a temporal one—when do we start?[17]

Joshua Lederberg, the molecular geneticist and Nobel laureate, also believes that eugenics is the field of the future. At the Ciba symposium he commented, ". . . it would be incredible if we did not soon have the basis of developmental engineering technique to regulate, for example, the size of the human brain by prenatal or early postnatal intervention."

How do these men propose to decide which traits are "desirable" so that they can be propagated in future generations? Obviously, what we living humans consider desirable in our present social milieu may not at all be what future humans might want. The basis of eugenics rests on the untenable assumption of infinite wisdom and prescience on our part. It is surely the ultimate in hubris to presume to know what is good for *all* future generations. There is an absurdity in the narrow vision of these scientists, who seem not to be aware of their limitations. However brilliant in their scientific disciplines, they are clearly not qualified once they step outside them and into the realm of practical applications.

In part, these men have fallen into the trap that often stands between scientists and the realization of a mature social consciousness: reductionism, the operational form of modern scientific research. It requires that the system under consideration be first separated into its most minute components. The forest as a whole may thus pass unnoticed. Thus can gene replacement therapy, in the molecular geneticist's eyes, obscure the overall picture of human health problems. Reductionism can be a fine thing in the design of experiments, but it is no aid in choosing them.[18]

Symposiums like this do not occur frequently, which is all to the good. They accomplish nothing except to reveal the shallowness of social thought still to be found in the scientific community. (And this thinking is by no means confined to geneticists.) If scientific research builds cathedrals, then these are sand castles in a dream world where technology blithely expands forever and the laws of thermodynamics do not apply. On the other hand, there is a vital need for regular and frequent discussions dealing with the potential effects of science in the real world. Health and other real human concerns should be discussed by scientists and other professionals at a working, not an esoteric, level. Such sessions might not produce immediately tangible results, but something more important might emerge—a social consciousness.

Social awareness is not entirely new in scientific circles. In recent times its first major manifestation occurred when theoretical and experimental physicists realized what they had brought about in splitting the atom. But their subsequent exhortations and wringing of hands did not prevent the events of 1945. Later, physicists and other worried individuals did their best to formulate a sane nuclear policy, but with limited success. The Union of Concerned Scientists has for years informed legislators and the general public on the dangers of radioactive pollution and the inadequacy of established safety measures. Their influence, which is being felt more and more, is urgently needed as we face the proliferation of nuclear power plants in America. Ironically, even those who advocate the use of nuclear energy admit that there is no foolproof method available at present or in the forseeable future for disposing of spent radioactive wastes, which will remain a threat to life for many thousands of years. With nuclear power we buy a little more time for energy-intensive technology, but at the expense of the future.

The pollution of the earth by industrial products and wastes, as well as other forms of environmental mutilation, has aroused opposition on the part of public interest groups such as the Friends of the Earth, the Environmental Defense Fund and the Sierra Club. Their uphill struggle against the shortsighted destruction of the human habitat could use more support from scientists. Now a new type of pollution is on the horizon, that of microbial strains created directly by scientists themselves. This form of pollution may prove more intractable than chemicals and is potentially as dangerous as radioactivity. The Coalition for Responsible Genetic Research, founded by concerned biologists, has come forth forcefully at the birth of this new technology to combat its blind acceptance and automatic exploitation. The Coalition acts on the belief that scientists, having presented society with a tool that could radically transform it, bear a heavy responsibility to inform the public of the possible danger, benefits and irrelevancies of that tool. The purpose of the Coalition is to give society an opportunity to make a deliberate and informed choice concerning the applications of genetic engineering, rather than

being forced into the traditional public role of fatalistic submission to privately developed technologies. It is not surprising that this position has elicited powerful opposition from the
scientific establishment, anxious to preserve its freedom from
public accountability. The widespread absence of conscience
as a factor in the conduct of science today has been well illustrated by this conflict. Nevertheless, public-interest nuclei
are now appearing in many scientific societies; it is to be
hoped that these will catalyze the development of social consciousness on a broader scale among scientists.

Although still outside the mainstream, there is an increasing number of scientific groups which are deeply committed to
social involvement. Science for the People is a major example
—a frankly activist group with a defined social philosophy. Its
purpose is to inform people, especially those involved in hazardous occupations, about technological dangers, no matter in
what domain. The members seek out and analyze controversial issues in science and technology, and offer their
opinion and guidance concerning, for example, the safety of
the workplace. They have had notable success in bringing the
XYY chromosome case to public attention.[19]

The New Alchemy Institute represents a group of scientists
who are activists in a scientific sense. Their aim is to exploit
the sophistication of modern science to develop an ecologically
stable and fulfilling way of life as an alternative to the dise-
quilibrium of modern industrial society. The New Alchemists
believe that development of small-scale, decentralized technology, particularly in food production, is the most promising
route to stable societies that can live in harmony with nature.
Using family-sized terrestrial capsules and other innovations
for food production, they have shown that efficient, small-
scale, highly intensive farming can be successful under a wide
variety of climatic conditions. Although the procedures are
not energy-intensive, this type of farming does not entail a
return to old agrarian ways—quite the contrary. The methods
are so efficient that a family can supply all its year-round
nutritional needs as a collateral activity. As an additional advantage, small-scale local production of the necessities of life is
expected to result in more emphasis on community life, less

fragmentation among individuals, and hence a greater feeling of responsibility and purpose. The New Alchemists recognize that new methods of food production will not solve all the world's problems, but they hope that by showing that sophisticated small-scale agricultural technology is feasible, their ideas may prove seminal in other areas.[20]

Even within the scientific establishment there are nuclei of socially responsive individuals: the scientists who produced the GRAS report (discussed in Chapter 6) for example, or the group of mathematicians at MIT who produced the "Limits to Growth" study for the Club of Rome.[21] The importance of the MIT work can hardly be overestimated. It is a prime example of the use of research and high technology to pinpoint fundamental world problems; it has identified many areas where science could make vital contributions to their solution.

Science, and scientists, are among the precious resources of the human species. As it becomes more evident that we have evolved a way of life that is ultimately inconsistent with the laws of nature, it becomes more and more irresponsible for scientists to spend their time building intellectual cathedrals and contributing to technological empires, while the foundations for future disaster are laid. Science has more to offer than that. It could help to form another way of life that is both humanly satisfying and in equilibrium with nature. But, first, the scientist has to recognize the nature of the problem. He needs to go far beyond his area of specialization. He has to consider the real significance of science—his work—in the world picture. Where is the present direction of research likely to lead? How may the results ultimately be applied? What impact would they have on the quality of life, judged in the overall context? Are there more important questions to be asked, and if so, how can his scientific expertise contribute to the answers? Conscience literally means "with knowledge," and surely that is how every scientist would wish to proceed.

References and Notes

Chapter 1. Dilemmas

1. André Cournand, *Science* 198 (1977): 699.
2. D. Dubarle, *Minerva* 1 (1963): 425.
3. J. Bronowski, *The Biological Revolution* (New York: Doubleday and Co., Inc., 1971) p. 305.
4. Jacques Monod, *Chance and Necessity* (New York: Vintage Books, 1972) and Ibid. p. 11.

Chapter 2. A Scientist Looks at Science: An Overview

1. DNA stands for "deoxyribonucleic acid," the macromolecule found in all cells—bacterial, plant, or animal—as well as in many viruses. The hereditary characteristics of these cells and viruses are contained in the DNA.
2. D. Greenberg, *The Politics of Pure Science* (New York: New American Library, 1967), gives an illuminating and incisive account of this period. See especially pp. 51–96.
3. Harvey Brooks discusses this in *The Government of Science* (Cambridge, Mass.: MIT Press, 1968).
4. O. T. Avery, C. McCleod, and M. McCarty, "Studies on Chemical Nature of Substance Inducing Transformation of Pneumococcal Types," *Journal of Experimental Medicine* 79 (1944): 137.
5. "Molecular Structure of Nucleic Acids," *Nature* 171 (1953):

158

737. After discussing the structure of DNA, Watson and Crick make the understatement of the century: "It has not escaped our notice that the specific pairing we have postulated immediately suggests a possible mechanism for copying the genetic material."

6. Once it was established that the sequence of the bases in DNA held the secret to genetics, the stage was set for all manner of theories and speculation. DNA became, and of course has continued to be, central to all biological phenomena. What had seemed to be disjointed findings suddenly crystalized into a coherent mass. The origin and function of other macromolecules, such as proteins and RNA (ribonucleic acid) seemed to fall into place. With time, the roles of small molecules such as sugars and amino acids became intelligible. An understanding of many of the functions of the cell, often referred to as a "bag of chemicals," was no longer a fanciful dream: the structure and function of many of its components could be defined in hard chemical terms.

7. *Cold Spring Harbor Symposium on Quantitative Experimental Biology* 18 (1953): 123. The program for the Cold Spring Harbor symposium, which took place in the summer of 1953, had been prepared in the previous year, as is the custom. The structure of DNA had been published in *Nature* only several months prior to the Cold Spring Harbor meeting, and the work would ordinarily not have been considered for inclusion in the program at that late date. However, the program committee felt the *Nature* paper by Watson and Crick to be so pregnant that it was placed on the regular program. Dr. Delbrick said in his opening remarks:

> Special mention should be made of a last-minute addition to the program, or rather, to the list of participants. The discovery of a structure for DNA proposed by Watson and Crick a few months ago, and the obvious suggestions arising from this structure concerning replication, seemed of such relevance to many of the questions to be discussed at this meeting that we thought it worthwhile to circulate copies of three letters to *Nature* concerning this structure among the participants before the meeting, and to ask Dr. Watson to be present at the meeting.

8. Much of the early research that led to the isolation of the restriction enzymes, which are central to recombinant DNA technology, was carried out by Werner Arber and his associates in Geneva, Switzerland. Some of Professor Arber's work is summarized in an article written for the *Annual Reviews of Biochemistry* 38 (1969): 467. Annual Reviews, Inc., Palo Alto, California, 1969. See also *Science* 202 (1978): 1069.

9. Actually, the recombinant DNA drama had been anticipated early on by Robert Pollack, who in 1971 had made known to Paul Berg his concern over the insertion of the monkey cancer virus SV40 DNA into E. coli. In commenting in general on the dangers of virus work Dr. Pollack, together with Dr. Joseph Sambrook, had the following to say to prospective virologists:

> Finally, you ought to ask yourself if the experiment needs to be done, rather than if it ought to be done, or if it can be done. If it is dangerous, or wrong, or both and if it doesn't need to be done, just don't do it. This is not censorship. You must accept a physician's responsibility if, by free choice, you work within these classes of experiments. Quoted from John Lear, *Recombinant DNA, The Untold Story* (New York: Crown, 1978), p. 27.

The Gordon Conference on Nucleic Acids in 1973 was the scene of much excitement in what was later to become the recombinant DNA controversy. At that conference a letter was written to the president of the National Academy of Sciences, setting forth the concerns of the molecular biologists present:

> We are writing to you, on behalf of a number of scientists, to communicate a matter of deep concern. Several of the scientific reports presented at this year's Gordon Research Conference on Nucleic Acids (June 11–15, 1973, New Hampton, New Hampshire) indicated that we presently have the technical ability to join together, covalently, DNA molecules from diverse sources. Scientific developments over the past two years make it both reasonable and convenient to generate overlapping sequence homologies at the termini of different DNA molecules. The sequence homologies can then be used to combine the molecules by Watson-Crick hydrogen bonding. Application of existing methods permits subsequent covalent linkage of such molecules. This technique could be used, for example, to combine DNA from animal viruses with bacterial DNA, or DNAs of different viral origin might be so joined. In this way new kinds of hybrid plasmids or viruses, with biological activity of unpredictable nature, may eventually be created. These experiments offer exciting and interesting potential both for advancing knowledge of fundamental biological-pro-

cesses and for alleviation of human health problems.

Certain such hybrid molecules may prove hazardous to laboratory workers and to the public. Although no hazard has yet been established, prudence suggests that the potential hazard be seriously considered.

A majority of those attending the Conference voted to communicate their concern in this matter to you and to the President of the Institute of Medicine (to whom this letter is also being sent). The conferees suggested that the Academies establish a study committee to consider this problem and to recommend specific actions or guidelines, should that seem appropriate. Related problems such as the risks involved in current large-scale preparation of animal viruses might also be considered. *Science* 181, (1973) 1114.

It is reported that the editor of *Science* concerned about the eventual effects of this letter on the scientific community, asked Dr. Maxime Singer, "Do you really want to do this?" The editor was not too anxious for speedy publication; the letter was published about two months later.

Not long after, the Berg Committee was formed under the auspices of the National Academy of Sciences. Its report was published in *Science* in July 1974 (p. 303). The first Asilomar Conference followed. For interesting accounts of these conferences, as well as the story of the recombinant DNA controversy, see Nicholas Wade, *The Ultimate Experiment* (New York: Wade, Walker, 1977); Michael Rogers, *Biohazards* (New York: Knopf, 1977); June Goodfield, *Playing God* (New York: Random House, 1977); J. Rifkin and T. Howard, *Who Should Play God* (New York: Delacorte, 1977); and Lear, *Recombinant DNA, The Untold Story.*

10. The concept of the genetic code can be described as follows: The DNA molecule contains a linear sequence of the four bases adenine, thymine, guanine, and cytosine. A word in DNA contains three bases, which may appear in any combination. These three-base sequences determine which amino acids will appear in the proteins specified by the DNA; a protein is a molecule containing many amino acids joined permanently together. *Three* bases in DNA correspond to *one* amino acid in the protein; the sequence of bases in DNA is co-linear with the sequence of amino acids in the protein. A change of sequence in these three-letter words leads to a corresponding change in the amino acid of the protein; a new protein is the result. The sequence can be changed in a variety of ways.

For a more detailed account of some of the ideas described here, see C. Grobstein, "The Recombinant DNA Debate," *Scientific American* 237 (1977): 22.

11. Nicholas Wade of *Science* Magazine has done a magnificent job of describing in three articles the details surrounding the activities of these teams. See *Science* 200 (1978): 279, 411, 510.

12. J. D. Watson, *The Double Helix* (N.Y.: Atheneum, 1968).

13. Philip Siekevitz has written a critique of prizes and their effects on scientific research. His commentary was instigated by the initiation of prizes by General Motors for cancer research. Professor Siekevitz notes:

I am still being continually surprised that seemingly the scientific community takes all these prizes seriously, without any thought as to what is implied in their giving and receiving. What is implied is that scientists primarily work for the prize, for the money, not for the curiosity, for the excitement, for the scientific knowledge. What is implied is that gains in scientific knowledge are primarily achieved by very exceptional individuals, without any input from the tens whose work in the same field has led to a temporarily final formulation. The organizers of these prizes apparently have no idea as to how advancements in scientific understanding are obtained, do not understand the "communitas" of the scientific community, that it is not made up of individually isolated scientists, but is a true community composed of interacting individuals. Thus, the General Motors prizes are to be given for recent published work, as if all that has gone before can be ignored, and only the final step is to be recognized by the huge monetary award. Is it not about time that the scientific community try to put a stop to the prostitution of its procedures and goals and aims by a money-oriented segment of our society that recognizes no other achievements than those which can be monetarily inspired? *Science* 202 (1978): 574.

My personal view about the Nobel Prize is that it is an exceedingly dangerous device with connoted privileges that extend quite far. I refer to virtual limitless power that an individual attains within his institution and among his colleagues; a power and authority that often transcends reasonable limits; a power that is sometimes destructive, especially when used against a less powerful adversary. Furthermore, the power is exercised within prestigious societies such as the National Academy of Sciences, which itself has vast authority in a variety of governmental agencies and activities. The web of power is clear. Nor does it end here. Nobel laureates are frequently sought as highly lucrative consultants by commercial interests such as drug corporations.

14. This subject is discussed at length in Ina Spiegel-Rosing and Derek de Solla Price, eds., *Science and Technology* (Beverly Hills, Calif.: Sage, 1977), chap. 4, p. 117. And in an article entitled "Hubris in Science?" appearing in *Science* 200 (1978): 1419, Lewis Thomas discusses various aspects of modern science. In discussing the important subject of communication between and among scientists Dr. Thomas states:

> The most surprising thing about the [communication] system is that it seems to be functioning with considerable accuracy and reliability. It is also surprising that there is so much openness and candor. It used to be thought that scientists tend to be rather secretive, hiding their data away from each other in order to be sure of priority for the published manuscripts; but these days it seems as though they are all telling each other everything they know, by telephone, and as soon as they know it.

I agree that there is considerable exchange of facts and views, but in my experience this is limited to the "in" people.

15. B. L. R. Smith and Joseph J. Karlesky, *The State of Academic Science* (New York: Change Magazine Press, 1977), p. 245. In their analysis of the peer-review problem, Smith and Karlesky conclude:

> The challenges to the peer review process may in any event arise more sharply in the future. With limited funds, there will be more "losers" to complain about unfair treatment. As one ingenuous faculty member observed during a site visit interview: "Peer review used to be fine. I'm all for peer review as it worked before. But now they are turning down everything, and good scientists are not being supported. Something is wrong." We believe that it will be particularly important to protect the integrity of the peer review system in those areas where it has worked well—in the traditional disciplines where scientists can evaluate the merits of competing proposals with a high degree of reliability.

16. Ibid., p. 37.

17. In 1978, 47,500 proposals were submitted to various federal granting agencies, involving the equivalent of about 6600 academic persons assuming each spends one-half of his time on research. The process is remarkably inefficient, especially if we note that three quarters of the proposals are rejected. See Editorial by A. C. Leopold, *Science* 203 (1979).

18. *Bulletin of the Atomic Scientists* 4 (1948): 69–72.

19. David Baltimore, quoted in Rogers, *Biohazard*, p. 52.

20. Many problems facing modern science were thought to be of such extreme importance that they were discussed at a Ciba Foundation symposium, the proceedings of which were published in *Science and Civilization* (North Holland: Elsevier, Excerpta Medica, 1973). The problem was defined in part by the opening remarks of the chairman, H. Bloch:

> The question of the worthwhileness of scientific activities was almost taboo when it was mentioned at that meeting in 1967. But now powerful voices from scientific as well as non-scientific quarters are saying that scientists should become more socially responsible, that science should be tolerated and supported only as long as its results are socially relevant, and that science must be constitutionalized and controlled if it is not to destroy our civilization. It is said that the scientist's lack of values has left him helpless to prevent science from being used for exploitation and destruction. And in the minds of many, science, all the way from nuclear physics and engineering to biology and medicine, has become a most dangerous evil.

In continuing his remarks Bloch posed a number of important questions about the scientific enterprise. He noted:

> These are new questions. They are indeed difficult. They reflect an ill-defined but strongly felt malaise about the role of science in society and about the scientist's responsibility for the technological application of the results of his work. It behoves us to recognize this malaise. Its aetiology is clear: it lies in the contribution of science to the deterioration of our world—or rather in the uncontrolled application of scientific technology that leads to the now well-known problems of environmental pollution, the use of science for war and destruction, and the social implications of the by-products and side effects of medical progress—and in fact that science and technology have failed in many people's view to make

our lives happier and more meaningful.

In the same symposium, E. Shils, a professor of sociology and social thought, pinpointed the problem:

What has been articulated, alongside of and often in the same minds as great awe and admiration, is a repugnance towards scientists for having, in collaboration with businessmen and the military, polluted the atmosphere and the waters, eroded and scarred the earth's surface, damaged its plant and insect life (p. 36).

21. Harry S. Hall, in "Scientists and Politicians," *Bulletin of the Atomic Scientists* 12 (1956): 46, discusses the congressional hearings, giving references to the original reports.

22. V. Bush, *Endless Horizons* (Washington, D.C.: Public Affairs Press, 1946), quoted by B. Glass, *Science* 171 (1971): 23.

23. Greenberg, *The Politics of Pure Science*, pp. 151–69, discusses the political maneuvering and the ensconcement of scientists in government associated with the emergence of Big Science since World War II.

24. *Report of the Presidents Biomedical Research Panel* (Washington, D.C.: U.S. Dept. of Health, Education, and Welfare, 1976). Publication No. (OS) 76-500.

25. E. L. Hess, *Federation Proceedings* 36 (1977): 2647. Dr. Hess was the executive director of the Publications Committee of the Federation Proceedings.

26. A. M. Silverstein, *Federation Proceedings* 37 (1978): 105. Dr. Silverstein was a congressional science fellow with the Senate Health Subcommittee.

27. Daniel S. Greenberg, "Report of the President's Biomedical Panel," *New England Journal of Medicine* 294 (1976): 1245. He observed:

That the panelists quaffed deeply of the ideology that holds basic research an undiluted good and an indispensable ingredient of health care is to be seen from its endorsement of a report prepared by a supporting subpanel chaired by Lewis Thomas, president of Memorial Sloan-Kettering Cancer Center:

Human beings have within reach the capacity to control or prevent human diseases. . . . There do not appear to be any impenetrable, incomprehensible diseases. . . . What is needed now is some sort of settling down to the long haul. . . . Most of all, the scientific enterprise needs stability and predicability. It does not require growth and expansion at the rate achieved in the 1950s and 1960s, but it cannot survive being turned on and off, nor will it succeed if held at a standstill without any opportunity for growth.

The response to this unqualified optimism was the assertion, "The Panel subscribes to this view of the future."

I doubt, however, that the same can be said of the U.S. Congress, the office of the HEW assistant secretary for health, or the Office of Management and Budget.

28. See note 20. See also *Man and His Future* (London: Churchill, 1963); and W. Fuller, ed., *The Biological Revolution* (Garden City, N.Y.: Doubleday Anchor, 1970).
29. Philip Handler, "Science's Continuing Role," *BioScience* 20 (1970): 1101.
30. Smith and Karlesky, *The State of Academic Science.*
31. The remarks of J. D. Watson are illuminating on this point. See N. Wade, "Gene-splicing Rules: Another Round of Debate," *Science* 199 (1978): 31, in which he quotes Professor Watson.
32. The Americans for Democratic Action adopted a resolution that according to a report in *Science* 197 (1977): 348, averred that strict societal control of science has preceded such excesses as Lysenkoism and some of the inhuman practices in Nazi Germany. Unfortunately, Walter Sullivan of the *New York Times* (July 31, 1977) took up the same theme at the beginning of his article, but recanted somewhat about two thirds of the way through it.
33. Bentley Glass, "Science: Endless Horizons or Golden Age?" *Science* 171 (1971): 26.
34. See for example H. S. Schumacher, *Small Is Beautiful* (New York: Perennial Library 1973), p. 171.
35. John Todd, *Journal of the New Alchemists* 3 (1976): 54.
36. Theodore Roszak, *Where the Wasteland Ends* (Garden City, N.Y.: Doubleday Anchor, 1973).

Chapter 3. Gene-Splicing

1. E. P. Odum, in *Fundamentals of Ecology* (Philadelphia: Saunders, 1971), p. 34, discusses feedback phenomena in various biological systems.
2. The network of interactions among cellular components resembles in some ways the electrical circuitry of a radio. Just as there are many electrical components in one circuit connected so that one

function (i.e., sound) is eventually realized, so there are a variety of chemicals and chemical interactions involved in any given metabolic pathway (circuit) that will give the desired end result. One metabolic pathway will produce a chemical that is essential for another pathway, which in turn yields a product required by another pathway, and so on almost *ad infinitum*.

3. In general the cell contains structural molecules and enzyme molecules. Enzymes have the capacity to do work, which in chemical terms means that they can join two (or more) molecules or they can degrade a larger molecule into its components. The synthesis of DNA requires a large number of enzymes: those that produce the precursor molecules in addition to those that join these precursors (of which there are four types) into a large DNA molecule. Not all the proteins required for DNA synthesis are enzymes. Some of the proteins are needed to put the parental DNA into a proper configuration for templating.

4. We may think of each of the four different groups in DNA as representing a "letter" of a word in a sentence of English; a number of these "words" gives rise to a gene whose counterpart in ordinary language is a sentence. Different permutations of the letters and words give rise to different genes.

Since the information for a protein is contained in DNA, it is necessary to make the four DNA groups congruent with twenty amino acids present in most proteins. It was the theoretical physicist George Gamow who in 1955 deduced that three of the four DNA "letters" would be required to "spell" (i.e., determine a single amino acid in the protein). This is the mathematical basis of the genetic code. A typical gene has about one thousand nucleotide "letters"; since three determine one amino acid, there are about 333 amino acid in a typical protein; in general, one gene carries the information for one protein. See Marshall Nirenberg, "The Genetic Code" *Scientific American* 208 (1963): 80. Number 3.

5. As a tool for molecular biologists, restriction enzymes are extremely powerful because of a unique property that enables these enzymes to cleave any DNA at a specific site, which is determined by the sequence of the four groups. Moreover, the enzyme cleaves the DNA in a specific way, creating "sticky" ends of the DNA; thus

$$—3 \quad 2 \quad 1 \quad 4 \quad 3—$$
$$—3' \quad 2' \quad 1' \quad 4' \quad 3—$$

represents a hypothetical site recognized by the restriction enzyme. The primed numbers represent groups complementary to the unprimed numbers. Cleavage occurs as follows:

-----3 + 2 1 4 3-----------
--------3' 2' 1' 4' + 3'------

When DNA molecules that have been cleaved by restriction enzymes are mixed, their complementary ends (i.e., their other "sticky" ends) cause the fragments to adhere to each other. This is another example of the principle of complementarity discussed in the text. The power of the technique lies in the fact the *different* DNAs have the *same* complementary ends when they are created by one restriction enzyme. Therefore, animal DNA can be joined to viral or bacterial DNA because their cleaved ends are complementary.

6. See discussions by Paul Berg and Maxime Singer, and by Robert L. Sinsheimer, *Federation Proceedings* 35 (1976): 2540, 2542; Paul Berg, *American Society for Microbiology News* 42 (1976): 273; E. Chargaff and F. Simring, *Science* 192 (1976): 938, 960; F. J. Dyson, *Science* 193 (1976): 6; N. Wade, *Science* 194 (1976): 303; B. D. Davis and J. Weizenbaum, *Wall Street Journal*, April 5, 1978; and P. Handler, B. Davis et al., *Chemical and Engineering News*, May 30, 1977, and April 17 and May 9, 1978.

7. Roger Noll and P. A. Thomas, "The Economic Implications of Regulation by Expertise: The Guidelines for Recombinant DNA Research," in *Research with Recombinant DNA* (Washington, D.C.: National Academy of Sciences, 1977): 262. Noll and Thomas make the following remarks regarding benefit-risk analysis:

The primary sin of commission in the debate about recombinant DNA research and the desirability of the guidelines has been the simplistic and largely inappropriate use of benefit-risk analysis to evaluate the research. In debating the value of their research in terms of benefits and risks, the molecular biologists have overstepped the bounds of their technical expertise, with the result that crucial aspects of a valid benefit-risk analysis are omitted or incorrectly treated in the discussion. The following are but a few examples to illustrate the point.

Among the issues missing from the benefit discussion are: (1) an assessment of the probability that any of these possibilities will be commercially attractive, (2) an estimate of the amount of time it will take for knowledge to be sufficient to make these objectives technically possible, (3) an estimate of the costs of the research that must be done before society will know whether commercial use of DNA recombination is worthwhile, and (4) the design of a comprehensive program of research that would contribute to the achievement of these public health and agricultural objectives.

Of course, the potential benefits of recombinant DNA research may also be reachable by other means. A precise statement of the benefits that might accrue from recombinant DNA research is that it may contribute to disease treatment, food production, and several other objectives, just as other lines of research may also make contributions in the same areas. A valid benefit-risk analysis would estimate the extent to which some expenditures on recombinant DNA research would increase the chance that society will capture these benefits for a given total expenditure on all paths to the same ends. For example, is a better way to reduce the death rate from cancer to seek cures for viral cancer through recombinant DNA research, or to expand research on environmental causes of cancer? Or, if in the long run insulin supplies are likely to run short, how should emphasis be divided among recombinant DNA studies, research on other synthetic processes, or expansion of supplies from animals?

8. U.S. House of Representatives, *Report of the Subcommittee on Science, Research and Technology of the Committee on Science and Technology* (Washington, D.C.: U.S. Government Printing Office, December 1976), Appendix 14, p. 251. Relevant excerpts from Dr. Davis's testimony follow.

I conclude that in the kinds of experiments now permitted (which exclude the introduction of known gene for a potent toxin or a known tumor virus) the danger of a significant laboratory infection is vanishingly small compared with the dangers encountered every day by medical microbiologists working with virulent pathogens. And such dangers must ultimately be balanced against the potential benefits, both practical and cultural.

Vigilance concerning new knowledge that might someday be misused is, to me, a threat to freedom of inquiry, and I believe a threat to human welfare. If we are entering dangerous territory in exploring recombinant DNA, we may enter even more dangerous territory if we start to limit inquiry on the basis of our incapacity to foresee its consequences.

9. Robin Holliday, "Should Genetic Engineers be Contained?" *New Scientist*, February 17, 1977, p. 399. An excerpt from this article follows:

I will not initially assign particular probabilities to each event, but simply refer to them as P_1, P_2, P_3, . . . etc. Similarly, I call the possible deleterious consequences C_1, C_2, C_3, . . . etc." [Dr. Holliday then enumerates the possible accidents (or events) and continues:]

The scenarios produce the following formulae: First pathway of infection (no specific type of pathogen):

$$P_1 \times P_2 \times P_3 \times P_3 \times P_5 \times P_6 \qquad = C_1 \text{ (death by infection)}$$

$$P_1 \times P_2 \times P_3 \times P_4 \times P_5 \times P_6 \times P_7 \times P_8$$
$$= C_2 \text{ resistant carrier, causing others' death)}$$

$$P_1 \times P_2 \times P_3 \times P_4 \times P_5 \times P_6 \times P_7 \times P_8 \times P_9$$
$$= C_3 \text{ (epidemic)}$$

Second Pathway of infection (induction of cancer):

$$P_1 \times P_2 \times P_3 \times P_4 \times P_5 \times P_{10} . \qquad = C_1 \text{ (death by cancer)}$$

$$P_1 \times P_2 \times P_3 \times P_4 \times P_5 \times P_{10} \times P_{11} \qquad = C_4 \text{ (cancer epidemic)}$$

[After assigning what he thinks are reasonable values to the various probabilities, P_1–P_{11}, he notes:]

We can now tot up the probabilities of the various deleterious consequences:

$$C_1 = 10^{-2} \times 10^{-1} \times 10^{-4} \times 10^{-3} \times 10^{-1} \qquad = 10^{-11}$$
$$C_2 = 10^{-11} \times 10^{-1} \times 10^{-1} \qquad = 10^{-13}$$
$$C_3 = 10^{-13} \times 10^{-1} \qquad = 10^{-14}$$
$$C_4 = 10^{-10} \times 10^{-2} \times 10^{-2} \qquad = 10^{-14}$$

Thus the probability of one individual dying of cancer is one in a hundred billion, C_1; the probability of a second individual dying is one in ten trillion, C_2; the probability of an epidemic of infections is one in a hundred trillion, C_3; the probability of a cancer epidemic is one in a hundred trillion, C_4.

10. Testimony of Arthur Schwartz, in U.S. Senate, *Hearings of the Senate Subcommittee on Science, Technology and Space of the Committee on Commerce, Science and Transportation* (Washington, D.C.: U.S. Government Printing Office, November 1977), pp. 307–309.

11. Union of Concerned Scientists, "The Risks of Nuclear Power Reactors" (Review of the NRC Reactor Safety Study Wash-1400, Washington, D.C., August 1977).

12. Ibid.

13. This was reported at the Workshop on Studies for Assessment of Potential Risks Associated with Recombinant DNA Experimentation, held at Falmouth, Mass., June 21–22, 1977. This workshop was sponsored by the National Institute of Allergy and Infectious Diseases.

14. Janet L. Hopson, "Recombinant Lab for DNA and My 95 Days in It," *Smithsonian*, June 1977, p. 55.

15. U.S. Senate, *Joint Hearing Before the Subcommittee on Health of the Committee on Labor and Public Welfare and the Subcommittee on Administrative Practice and Procedure of the Committee on the Judiciary* (Washington, D.C.: U.S. Government Printing Office, September 1976), p. 102.

16. W. Szybalski, "Safety of Coliphage Lambda Vectors Carrying Foreign Genes" in *Recombinant Molecules: Impact on Science and Society*, R. F. Beers, Jr. and E. G. Bassett eds. (New York: Raven Press, 1977), p. 138.

17. Seymour Lederberg, "The Least Hazardous Course," *Recombinant Molecules: Impact on Science and Society*, R. F. Beers, Jr. and E. G. Bassett eds. (New York: Raven Press, 1977) p. 485.

18. R. Holliday, "Genetic Engineers," p. 400.

19. Joshua Lederberg, *Report of the Subcommittee on Science*, Appendix 11, p. 233.

20. Quoted in Rogers, *Biohazard*, p. 81.

21. Barry Commoner discusses this in *The Closing Circle* (New York: Knopf, 1977), pp. 268 et seq.

Chapter 4. The Hazards of Success

1. The first Asilomar Conference was held at Pacific Grove, California, in February 1975. See note 9, chapter 2, and notes 13 and 14, chapter 7, for further details.

2. Only after publicity about the DNA controversy was any serious effort devoted to the study of other possible host cells and vectors. Between 1975 and 1978, about $2 million had been allocated by the National Institutes of Health for such research. This sum is only a small fraction of the money spent on research using the E. coli, which lives everywhere and virtually on, or in, all living things.

3. This criticism was contained in the U.S. Senate, *Report of Oversight Hearings by the Subcommittee on Science, Technology and Space of the Committee on Commerce, Science and Transportation* (Washington, D.C.: U.S. Government Printing Office, August 1978).

4. On September 15, 1978, a public hearing, advertised in the *Federal Register*, was held by officials of HEW in order to consider the relaxed NIH guidelines and see whether they were suitable for administration by HEW. The participants at this hearing were chosen from a broader range of interested parties than that of previous hearings held by the National Institutes of Health. The original guidelines were binding on scientists doing NIH-supported work. They did not apply to industry. To bring industry under the guidelines, the secretary of HEW asked FDA and EPA to take action under their legal authorities.

5. Articles on the commercialization of recombinant DNA technology are in *Chemical and Engineering News*, July 18, 1977, June 19, 1978, and September 11, 1978; *Business Week*, December 12, 1977, and January 17, 1977; *Boston Globe*, June 11, 1978; *New York Times*, June 9, 1978; *New York Times*, Jan. 27, 1980; *The Economist*, p. 71 Jan. 1980; *New York Times*, Feb. 3, 1980.

Many small companies have been formed in addition to Genentech. These include: Biogen (Swiss), Genex (U.S.), Cetus (U.S.), Hybritech (U.S.). Large corporations (Hoffman-LaRoche, Dupont, Upjohn, Chevron) have bought major interest in small companies or are starting their own research.

6. Articles in *Science* 202 (1978): 724; *Business Week*, September 8, 1977; *Washington Post*, July 11, 1979; *Nature* 494, 385 (1979): 278.

7. *Nature* 274 (1978): 2. The Patent and Trademark Office had originally turned down the application for the patenting of streptomyces vellosus. The U.S. Court of Customs and Patent Appeals reversed the decision and allowed the patent. In late June 1978, the U.S. Supreme Court ordered the appeals court to review its decision.

8. A study of nuclear risks was carried out under the auspices of the Atomic Energy Commission and was first released in 1974. This report, known as the Rasmussen Report, ostensibly a comprehensive study of nuclear safety, nevertheless made many assumptions questioned by the Union of Concerned Scientists in their report "A Review of the National Regulatory Commission Reactor Safety Study (Wash-1400)," published in 1977. Early in 1979 it was finally admitted by government officials that the Rasmussen Report had serious flaws in it, which considerably weakened the case for nuclear "safety." It was also admitted that the report did not constitute a set of guidelines, although originally this was the impression created for public consumption.

Another "incident" (not to say coverup) relating to nuclear safety has recently come to light (*International Herald Tribune*, January 30,

1979, p. 3). A study of the effects of low-level radiation on workers in nuclear plants at Hanford, Washington, and Oak Ridge, Tennessee, was terminated by the Atomic Energy Commission in 1974, after thirteen years of research, because the study showed that dangerous effects were possible. Most experts agree that such radiation results in a higher incidence of cancer. The scientist responsible for the research, Dr. Thomas Mancuso, said that "the AEC set out to fund a political study with guaranteed negative findings. When they found out that their political purpose had collapsed, they dumped me." Recently a research worker with the National Institute for Occupational Safety and Health stated, referring to low-level radiation, that "we are uncovering some significant biological effects often of alarming proportions." He indicated that low-level radiation may cause cancer.

9. The report of the federal Interagency Committee on Recombinant DNA Research states: "As of the summer of 1977, there were an estimated 150 recombinant DNA projects under way in Europe, 300 in the United States, and perhaps 20-25 altogether in Canada Australia, Japan, and the Soviet Union. All are being conducted under some form of safety practices and procedures." As of December 1979 the NIH was sponsoring over 700 projects.

10. Lear, *Recombinant DNA, p. 246.*

11. *New York Times,* August 5, 1978.

12. Gertrud-Barna-Lloyd draws attention to the distinction she believes should be made between the terms *pollutants, industrial chemicals,* and *life-style factors,* all of which are frequently subsumed under the rubric *environmental.* Examples of life-style factors are sunshine, diet, and cigarettes; the latter, she says, should be targets for cancer prevention just as are "chemical factors." *Science* 202 (1978): 469.

The epilemiologist E. L. Wynder notes: "Epidemiologic studies on cancer have shown that most human cancers are environmentally caused, that is to say, are man made, and are, therefore, preventable. They are generally related to habits such as smoking, poor nutrition, excessive alcohol consumption as well as to certain occupational exposures."

13. Richard Doll, "Strategy for Detection of Cancer Hazards to Man," *Nature* 265 (1977): 589.

14. Irving R. Johnson, *Research with Recombinant DNA* (Washington, D.C.: National Academy of Sciences, 1977), p. 156.

15. Ruth Hubbard, *Research with Recombinant* DNA p. 168.

16. World Health Organization report quoted in the *New York*

Times, October 25, 1977.

17. *Science* 199 (1978): 162.

18. *New York Times,* December 11, 1977, and January 3, 1978.

19. W. Ophuls, *Ecology and the Politics of Scarcity* (San Francisco: Freeman, 1977), p. 46.

Chapter 5. Science as Technology, and Vice Versa

1. Statement of Dr. Philip Handler, in U.S. Senate, *Hearings Before the Subcommittee on Science, Technology and Space of the Committee on Commerce, Science and Transportation,* 95th Cong., 1st sess. (Washington, D.C.: U.S. Government Printing Office, 1977), p. 4.

2. Letter from David Perlman to the editor, *Science* 198 (1977): 782.

3. At that time I wrote the following letter (never published) to the editor of *Science:*

David Perlman (Letter 25 November 1977) has raised an interesting and fundamental point concerning the dissemination of scientific knowledge and how this relates to the political process. I agree with Perlman when he suggests that the propriety of Dr. Handler's testimony, before a Senate hearing (November 2, 1977), in which unpublished recombinant DNA results were cited, should be brought into question and discussed publicly. There are at least two other significant instances in which unpublished data concerning recombinant DNA appeared in the public press before appearing in a scientific journal. The first is the recent "Falmouth" affair, in which results and opinions expressed at a closed scientific conference on risk assessment of recombinant DNA technology were summarized by the Chairman in a letter to Dr. Donald S. Fredrickson, Director of NIH. Not only did the letter contain unpublished data but it was sent without prior approval of the conferees. Regardless of the content of the letter, the action was inappropriate and was so described by some members of that conference. The letter was interpreted by many, rightly or wrongly, as giving a "clean bill of health" to recombinant DNA technology. Indeed the Editor of Science quoted material from this letter in an editorial (August 19, 1977). Articles in the public press (Walter Sullivan, *New York Times*) also publicized the Falmouth meeting, again stating that the dangers of recombinant DNA technology had been exaggerated. Thus the letter has given rise to a public debate based on unpublished data. The second case in point concerns the publicity based on an unpublished article dealing with recombinant DNA by Dr. Stanley Cohen. The unpublished results received a great deal of attention in the press (e.g., *Washington Post*, September 28, 1977), including interpretations which were not substantiated by the article itself. Regardless of the substance of the issue ("novel" organisms produced by recombinant DNA), the matter was presented to the public in an unfortunate way and without evidence of peer review.

Instances of this kind will decrease the credibility of scientists when the truth becomes known to the general public.

David Perlman should be thanked for his comment. I am happy to see that it was published in *Science*—it might have appeared instead in the *San Francisco Chronicle*.

4. *Chemical and Engineering News*, June 19, 1978, p. 4.

5. National Academy of Sciences, "Research with Recombinant DNA 156," an academy forum funded by the National Institutes of Health, 1977.

6. For an interesting discussion of this subject, see Greenberg, *The Politics of Pure Science*.

7. U.S. Department of Health, Education and Welfare, *Report of the President's Biomedical Research Panel* (Washington, D.C.: U.S. Government Printing Office, 1976), pp. 22–23.

8. For recent discussions of federal cutbacks in basic research, see *Science* 201 (1978): 330; and *Wall Street Journal*, June 26, 1978, p. 1.

9. Zinder report (*Report of Ad Hoc Committee to Conduct a Review of the Special Virus Cancer Program of the National Cancer Institute*, March 1974).

10. Biomedical Research Panel, Appendix A, p. 56.

11. For an interesting discussion of political pressures regarding the President's Biomedical Research Panel, see Silverstein, "Congressional Politics and Biomedical Science," *Federation Proceedings* 37 (1978): 105. This was in reply to Hess, "Wheel Spinning in Washington, Another Fizzle," *Federation Proceedings* 36 (1977): 2647.

12. Borek, "The Loneliness of the Original Investigator," *Nature* 264 (1976): 100.

13. A. Carl Leopold, Editorial, *Science* 203 (1979).

14. See, for example, T. Roszak, *Where the Waste Land Ends* (Garden City, N.Y.: Doubleday Anchor, 1973).

15. *New York Times*, October 25, 1977. The *Times* article based on a

report of the Expert Committee on Selection of Essential Drugs, which met in Geneva October 17-22, 1977.

Chapter 6. Rousseau Revisited

1. Quoted in Rogers, *Biohazard* p. 71.

2. *Washington Post,* May 24, 1977.

3. U.S. Senate, *Hearings Before the Subcommittee on Science, Technology and Space of the Committee on Commerce, Science and Transportation* (Washington, D.C.: U.S. Government Printing Office, November 1977), p. 182.

4. Ibid., pp. 208–10.

5. Ibid., p. 203.

6. Ibid., pp. 216, 217, 224.

7. Lear, *Recombinant DNA,* p. 266.

8. Robert K. Merton, "Behavior Patterns of Scientists," *American Scientist* 57 (1969): 1.

9. *Science* 200 (1978): 1438.

10. N. Wade, *Science* 200 (1978): 516.

11. U.S. Senate, *Staff Report to the Subcommittee on Administrative Practice and Procedures of the Committee on the Judiciary* (Washington, D.C.: U.S. Government Printing Office, December 1976), p. 36.

12. Ibid., p. 38.

13. See *Nature* 262 (1976): 636.

14. *Nature* 263 (1976): 538.

15. Ibid.

16. *Nature* 264 (1976): 309.

17. Ibid.

18. Ibid.

19. An excellent documentary, prepared for television by Thames Television, appeared on October 4, 1977. Research for this program was done by Jon Blair; the reporter was John Fielding; the director was Norman Fenton; and the producer was David Elstein. See also *Science* 192 (1976): 240.

20. Thames Television script, pt. 1, p. 6.

21. Ibid., pt. 3, p. 2.

22. Ibid., pt. 7, p. 7.

23. Ibid., pt. 3. p. 7.

The last chapter in the PBB incident is still not written. See Jane E. Brody in the *New York Times,* January 5, 1977 for new findings regarding afflicted people. The following headline appeared in the *Wall Street Journal* August 1, 1978: "Michigan Dairy Farmer Who in

1974 Found PBB in Cows Now Finds His New Herd Contaminated."

The article, by John R. Emshwiller, describes how persistent PBB has been despite persistent efforts at a clean-up. After washing barns and bins with high-pressure steam and recovering floors with concrete, PBB was still in evidence. "Uncontaminated" grazing land was nevertheless not usable. George Fries, a federal researcher, said: "Once it gets out there, it's impossible to clean up completely." This contamination problem points up the fact that there are unforseen consequences many years after an industrial chemical disaster. Can the Velsicol Corporation ever pay enough to those who have lost their health and their livelihood?

But industrial-chemical disasters are only one source of poisoning. A recent article in the *New York Times* (October 3, 1977) pointed out that one in four workers is exposed to hazards in the workplace. The 697-page report by National Institute of Occupational Safety and Health shows that hundreds of thousands of workers were exposed to substances believed to cause cancer or other fatal diseases; a large proportion of employees exposed to these substances had not been given medical tests to determine whether their health was threatened.

Another case in point is that of the contamination of the Hudson River and its estuaries with PCB (polychlorinated biphenyls). These chemicals were first manufactured in 1929 and have been used primarily as heat-transfer fluids and as insulators in heavy electrical equipment. The PCBs dumped into the waters by General Electric were discovered in the Hudson River in 1975. General Electric was sued by the Department of Environmental Conservation, and on September 8, 1976, a settlement was reached. The agreement stipulated that General Electric cease using PCBs by July 1977 and also contribute $3 million to the department for further work and monitoring (*Report of New York State Department of Environmental Conservation, Bureau of Water Research, Division of Pure Water*, July 1977). Revised estimates of the clean-up of the river go up to $150 million (*New York Times*, October 19, 1977). Worse, no one knows where to put the dredgings or even if it is wise to disturb the river in the first place; dredging could cause more contamination downstream. Yet General Electric is legally absolved, having paid a minuscule fine!

PCBs are only a few of the poisonous chemicals discharged into rivers by manufacturing companies. Recently, carbon tetrachloride was found in large quantities (one spill was seventy tons) in the Ohio and Kanawha rivers. The EPA discovered that four companies—

Allied Chemical, Diamond-Shamrock, FMC Corporation, and PPC Industries—produce carbon tetrachloride in plants above Huntingdon on the Ohio and Kanawha rivers. FMC had so many spills that the EPA brought suit and had one of its plants shut down because of the "record of past discharges and the deteriorated condition of its plant." EPA toxicologist Robert Tarcliff claims that the effects of carbon tetrachloride accumulate with repeated exposure. Tarcliff also believes that PCBs, barbiturates, and alcohol potentiate the toxic effects of carbon tetrachloride. Repeated small doses may also cause cancer. See *Science* 196 (1977): 632.

Chemicals can be ingested directly from water and also from eaten fish. The problem has become so serious that the New York State Department of Health issued an advisory on eating fish. The first paragraph reads:

Certain fishes of Lake Ontario and other waters of the state have accumulated subtle environmental contaminants. Fishes from some waters may contain levels of Mirex, PCBs (polychlorinated biphenyls) or mercury which exceed actionable or temporary tolerance levels established by the U.S. Food and Drug Administration (FDA) to protect consumers obtaining commercial fish in interstate commerce. Onondaga Lake is closed to fishing because of mercury contamination, and all fishing is prohibited in the Hudson River between Fort Edward and Troy Dam because of PCBs.

24. *Report of the Select Committee on GRAS Substances*, Bethesda, Maryland, Life Sciences Research Branch Office, Federation of American Societies for Experimental Biology, May 1977).

Chapter 7. From Truth to Power

1. The full story of the recombinant DNA drama is recounted in note 9 of chapter 2.

2. For an interesting commentary on truth and power, see K. Boulding, Editorial, *Science* 190 (1975).

3. J. D. Watson and F. H. C. Crick, "Molecular Structure of Nucleic Acids," *Nature* 171 (1953): 737.

4. E. Chargaff in E. Chargaff and J. N. Davidson eds., *The Nucleic Acids* (New York: Academic Press, 1955), chap. 10. See also R. Franklin and R. G. Gosling, "Molecular Configuration in Sodium

Thymonucleate," *Nature* 171 (1953): 740; and M. H. F. Wilkins, A. R. Stokes, and H. R. Wilson, "Molecular Structure of Deoxypentose Nucleic Acids," *Nature* 171 (1953): 738.

5. Watson, Crick, and Wilkins received the Nobel Prize for their work in 1962. Many have thought, and I concur, that Chargaff should have been similarly recognized. It is common knowledge that the equivalency of bases (adenine = thymine; guanine = cytosine) was due to Chargaff and his co-workers. "Chargaff's rules" provided the experimental basis of the base-pairing hypothesis used by Watson and Crick. In his book *Heraclitean Fire* (New York: Rockefeller University Press, 1978), Chargaff recounts his views on base-pairing and complementarity.

6. See Chargaff and Davidson, eds., *Nucleic Acids*, for a list of general early works on these acids.

7. L. Pauling and M. Delbruck, "The Nature of the Intermolecular Forces Operative in Biological Processes," *Science* 92 (1940): 77.

8. See note 4 of chapter 3 for a discussion of the genetic code.

9. The Ciba Foundation sponsored two symposiums, which resulted in two monographs. *Man and His Future* was published in 1963, and *Science and Civilization* in 1973. The quotation is from a contribution to the former symposium by Julian Huxley.

The British Society for Social Responsibility in Science sponsored a conference that resulted in *The Biological Revolution* (Garden City, N.Y.: Doubleday Anchor, 1978). The preface of this book pinpoints some of the problems we now face:

Advances in the treatment of disease, in understanding the processes of ageing, in the control of behaviour, in the harnessing of micro-organisms to produce particular chemicals, and in food production, have potentialities which will expose the limitations of the social and political structures which have evolved for the application of scientific and technical advances. The problem is not simply that we may not derive the maximum benefit from these advances but that if this new information is not correctly applied, our well-being or even the survival of our species is threatened. We are threatened physically as a species by the depletion of resources and poisoning of our environment because of unplanned technological developments. We are threatened socially as a species by the techniques for the control of behaviour which could dehumanize and could destroy creativity. As individuals, our value systems are being undermined as new knowledge on biological phenomena such as heredity and behaviour undermines the myths and dogmas at the heart of our ethical and religious beliefs.

10. P. Siekevitz, "Scientific Responsibility," *Nature* 227 (1970): 1301.

11. W. J. Broad, *Science* 201 (1978): 1195. In an article by H. F. Judson in the *New Yorker* magazine (December 11, 1978), Bragg is quoted as saying ". . . when it came to the Nobel Prize, I put every ounce of insight I could behind Wilkins getting it along with [Watson and Crick]."

12. See note 9 of chapter 2.

13. The details of actions leading to the NIH guidelines can be found in documents published by the Department of Health, Education, and Welfare. They are DHEW publications No. (NIH) 76-1138 and No. (NIH) 78-1139. The Environmental Impact Statement on the NIH guidelines for Research Involving Recombinant DNA molecules was published in two documents October 1977: DHEW publications No. (NIH) 1489 and No. (NIH) 1490.

14. "Statement on Recombinant DNA Research," United States Catholic Conference, Washington, D.C. 1977.

15. "National Academy of Sciences Forum on Research with Recombinant DNA," National Academy of Sciences, Washington, D.C., March 1977. See discussions by G. Wald, R. Hubbard, J. King, K. Mislow, and R. Sinsheimer.

16. There are two methods for the containment of bacteria within the laboratory; one is physical, the other biological. Each is defined by a level, depending on the assumed danger.

The physical levels are designated P1 to P4, and the biological levels EK1 to EK3, the EK referring to the K12 strain of Escherichia coli.

P1 is a standard microbiological laboratory with no special equipment or safety procedures. The work is done on open benches.

P2, the next level up, is not significantly different from P1. Eating and drinking in the work area are prohibited, as is mouth-pipetting —a mechanical pipetting device must be used instead.

P3 is the first level at which the laboratory requires special design and equipment. All work with organisms containing recombinant DNA molecules must be performed within biological safety cabinets. Access to the lab is through a double door.

P4 is the level of physical containment used at the army's

biological warfare laboratories. Entry and exit to the laboratory are tightly controlled.

The three levels of biological containment for recombinant DNA research are EK1, EK2, and EK3, defined as follows:

EK1 requires the use of the standard laboratory strain of E. coli, known as E. coli K12.

EK2 refers to host-cloning vehicle pairs in which the cloning vehicle has only 1 chance in 100 million of surviving outside a special medium.

EK3 is an EK2 system whose lack of survivability has been demonstrated by actual feeding tests in man and other animals.

17. *Chemical and Engineering News*, March 29, 1977.

18. In National Academy of Sciences Forum discussion.

19. *Nature* 265 (1977): 575.

20. *Nature* 263 (1976): 518.

21. For a more complete summary of federal recombinant DNA legislation, see DHEW publication No. (NIH) 78-1139.

22. Letter written by Dr. Roy Curtiss III to Director of NIH Donald Fredrickson on April 12, 1977. See U.S. Senate, *Hearings Before the Subcommittee on Science, Technology and Space of the Committee on Commerce, Science and Transportation*, Serial No. 95-52 (Washington, D.C.: U.S. Government Printing Office, November 1977), p. 48 (hereafter referred to as *Hearings on Science, Technology and Space*).

23. In reply to critics of the use of E. coli as a host for recombinant DNA, see *Science* 193 (1976): 215:

"You are . . . undoubtedly correct (in principle) that E. coli is the wrong microorganism," wrote DeWitt Stetten, NIH deputy director for science and chairman of the NIH recombinant DNA committee, in a letter of 6 October 1975 to a critic on this point. "Even at the Asilomar Conference, however," Stetten added, "I detected little interest on the part of the majority to table E. coli and begin again from scratch with some other organism. The enormous quantity of accumulated information about E. coli appeared to dictate that, despite its hazards, this was still the organism of first choice. . . ."

24. See testimony of M. Lappe, *Hearings on Science, Technology and Space*, p. 104.

25. S. Chang and S. Cohen, "In vivo site-specific genetic recombinations promoted by the EcoRI restriction Endomuclease" *Proceedings of the National Academy of Sciences* 74 (1977): 4811.

26. R. L. Sinsheimer, "On Coupling Inquiry and Wisdom," *Federation Proceedings* 35 (1976): 2540; "Potential Risks," National Academy of Sciences Symposium, March 1977, p. 74; N. Wade, "Recombinant DNA: A Critic Questions the Right to Free Inquiry," *Science* 194 (1976): 303; and *New York Times*, May 30, 1977. Sinsheimer concluded that "neither I nor anyone else can say that if the present committee guidelines are adoped, disaster will ensue. I will say, though, that in my judgment, if the guidelines are adopted and nothing untoward happens, we will owe this success far more to good fortune than to human wisdom." *Science* 192 (1976): 237.

27. *Washington Post*, September 28, 1977.

28. After Dr. Cohen's campaign, Senator Schmidt wrote, in a letter to *Science* 201 (1978): 106: "There have been recombinant processes occurring in nature since life began . . . ." Senator Kennedy commented, at the time he withdrew his bill, that Cohen's work "raises serious questions as to whether recombinant DNA can ever produce a 'novel' organism. . . ." See *Washington Post*, September 28, 1977, p. 1.

29. Quoted in Lear, *Recombinant DNA,* p. 242.

30. For details, see "Statement on Recombinant DNA Research," pt 2, M-1.

31. Letter written by Bruce Levin to Donald Fredrickson on July 29, 1977, quoted in *Hearings on Science, Technology and Space*, p. 131.

32. Letter written by Jonathan King and Richard Goldstein to Donald Fredrickson on August 22, 1977; ibid., p. 48.

33. The circumstances surrounding the scientists' lobby against recombinant DNA safety legislation have been described by Barbara Culliton in *Science* 199 (1978): 274.

34. See DHEW publication No. (NIH) 76-1138, p. 184.

35. Prominent individuals included W. D. McElroy, past president and board chairman of the American Association for the Advancement of Science; Edward E. David, Jr., president-elect of the American Association for the Advancement of Science; Emilio Q. Daddario, president of the American Association for the Advancement of Science; and W. D. Carey, executive officer, American Association for the Advancement of Science. See *Hearings on Science, Technology and Space*, p. 375.

36. Letter signed by Walter Gilbert, in *Science* 197 (1977): 208.

37. *Science* 197 (1977).

38. *Science* 199 (1978).

39. S. Levy and B. Marshall, *Recombinant DNA Technical Bulletin 2* (1979): 77-80; M. Chatigny, M. Hatch, H. Wolochow, T. Adler, J.

Hresko, J. Macher, and D. Besemer, *Recombinant DNA Bulletin 2* (1979): 62-67.

40. M. Israel, H. Chan, W. Rowe, and M. Martin, *Science* 203 (1979): 883-887; M. Chan, M. Israel, C. Garon, W. Rowe, and M. Martin, *Science* 203 (1979): 887-892.

41. B. H. Rosenberg and Lee Simon, *Nature* 282 (1979): 773; *Nature* 283 (1980): 796.

42. Lear, *Recombinant DNA*, pp. 244, 245.

43. Ibid., p. 8.

44. See *Hearings on Science, Technology and Space,* p. 17.

45. *Science,* 193 (1976): 216.

46. P. Boffey, *The Brain Bank of America* (New York: McGraw-Hill, 1975).

47. S. E. Luria, in an article entitled "The Goals of Science," *Bulletin of the Atomic Scientists* 33 (1977): 28, likens scientists to ancient burghers. He said:

For the enthusiastic scientist, the scientific enterprise is a monument to humanity's intellectual power and freedom—a modern equivalent of the great cathedrals that the burghers of the Middle Ages raised as monuments to their newly found sense of economic power and political freedom.

But, if science is a cathedral raised in praise of intellectual freedom, one must admit that too often, under the pressure of utilitarian society, the cathedral of science has come to look like one of those monasteries one sees in the French countryside, in which a modest church is almost hidden by a prosperous distillery. The sale of products becomes the justification for being allowed to pray to the Lord.

48. Dr. L. Cavalli-Sforza, at the International Symposium on Genetic Engineering held in Milan, March 1978.

Chapter 8. Freedom of Inquiry

1. U.S. Senate, *Joint Hearings Before the Subcommittee on Health of the Committee on Labor and Public Welfare and the Subcommittee on Administrative Practice and Procedure of the Committee on the Judiciary* (Washington, D.C.: U.S. Government Printing Office, September 22, 1976), p. 85; statement made by David Baltimore.

2. Baltimore, *Joint Hearings,* p. 73.

3. Paul Berg, in "Research with Recombinant DNA," a symposium sponsored by the National Academy of Sciences, March 1977, p. 73.

4. From statements made by Professor Green at the Asilomar Conference, 1975; quoted in Lear, *Recombinant DNA*.

5. *Chemical and Engineering News*, June 19, 1978, p. 4.

6. *Business Week*, December 12, 1977, p. 12, and January 17, 1977, p. 76 D; *Nature* 274 (1978): 2.

7. A. Cournand, *Science* 198 (1977): 699.

8. Smith and Karlesky, *The State of American Science*, p. 244.

9. Harold Green's remarks at the Asilomar Conference, 1975, Pacific Grove, California.

10. Harvey Brooks, W.R.D. Sewell et al, eds., *Modifying the Weather: A Social Assessment*, (Victoria, British Columbia: University of Victoria, 1973), p. 83.

11. Annual report of the Committee on Scientific Freedom and Responsibility, American Association for the Advancement of Science, 1977, p. 44.

12. Ibid.

13. Berg, "Research with Recombinant DNA," p. 278.

14. N. Wade, "Gene Splicing: Senate Bill Draws Charges of Lysenkoism," *Science* 197 (1977): 348.

15. Philip Handler, *Chemical and Engineering News*, April 17, 1978.

16. Bentley Glass, "Science: Endless Horizons or Golden Age?" *Science* 171 (1971): 23.

17 C. S. Lewis, *The Abolition of Man* (New York: Macmillan, 1973), p. 69.

Chapter 9. Conscience in Science

1. Lewis Mumford, *The Myth of the Machine*, 2 vols. (New York: Harcourt Brace Jovanovich, 1964, 1970); idem, *Technics and Civilization. Interpretations and Forecasts: 1922-1972* (New York: Harcourt Brace Jovanovich, n.d.); Robert Heilbroner, *An Inquiry into the Human Prospect* (New York: Norton, 1975); B. Barber, *Science and the Social Order* (New York: Collier, 1962); Robert K. Merton, *Social Theory and Social Structure* (New York: Free Press, 1949); idem, *The Sociology of Science: Theoretic and Empirical Investigations* (Chicago: University of Chicago Press, 1973); Jacques Ellul, *The Technological Society* (New York: Knopf 1964); J. Bronowski, *Science and Human Values* (New York: Harper & Row, 1956); Rene Dubos, *So Human an Animal* (New York: Scribner's, 1968); J. Haberer, *Politics and the Community of Sci-*

ence (New York: Van Nostrand Reinhold, 1969); and John M. Ziman, *Public Knowledge; An Essay Concerning the Social Dimension of Science* (New York: Cambridge University Press, 1968).

2. Robert Heilbroner, *An Inquiry into the Human Prospect*, (New York: W.W. Norton, 1974), p. 1.

3. Lear, *Recombinant DNA*, p. 246.

4. William Ophuls, *Ecology and the Politics of Scarcity*, p. 48.

5. Ibid., p. 55.

6. E. P. Odum, *Fundamentals of Ecology*, p. 515.

7. N. F. Jensen, "Limits to Growth in World Food Production," *Science* 201 (1978): 317. Professor Jenisen concludes:

> The dramatic increases in wheat yields that began in the mid-1930's in the United States will soon begin to level off. The favorable mix of genetics and technology that has characterized this era must build upon an ever higher yield base for the future. At the same time the residue of factors that can lower wheat yields includes a larger proportion of forces not easily shaped or controlled by man. An example is weather. The result is a natural yield ceiling that is already visible and that will impose a limit on future productivity growth.

8. D.H. Meadows, D.L. Meadows, J. Randers, and W.W. Behrens, III in *Limits to Growth* (New York: Potomac Associates, New American Library, 1972).

9. Secretary Califano in a speech given at the National Conference on Health Research Principles at the NIH, October 3, 1978. Transcript published in *Federation Proceedings* 37 no. 13 (November 1978): i.

10. David Baltimore, in "Research with Recombinant DNA," a forum of the National Academy of Sciences, March 1977, p. 238.

11. Jonathan Bechwith, *Annals of the New York Academy of Sciences*, 265, (1976): 46.

12. Jonathan Bechwith, "Research with Recombinant DNA," *Annals of the New York Academy of Sciences*, March 1977, pp. 243–244.

13. *International Herald Tribune*, February 3, 1979. Recently a federal advisory board agreed that the government should end a five-year ban on federally funded test-tube-baby projects. Although the thirteen-member board agreed that the government was faced with a "terrible dilemma," it nevertheless sanctioned a go-ahead. Evidently the board was pressured into this action because clinics in America and Europe were getting ready to provide the service. Al-

though Reverend Richard McCormick argued against the test-tube baby for study purposes, he agreed that it would be acceptable to carry out the procedure if the fertilized egg were implanted into the donor woman. The board agreed that embryos created for study purposes alone could not be used except by consent of the donors.

The fact that clinics all over the world were ready for action, regardless of government rule, is disturbing, as is the fact that the board saw "no ethical objections." It is clearly only a matter of time before genetic manipulations will occur, and with the laudable goal of providing "better" humans. Better?

14. The proceedings of the symposium were published in *Man and His Future*, (London: Churchill, 1963).

15. A partial list of participants includes Sir Julian Huxley, Colin Clark, Albert Szent-Gyorgyi, Hilary Koprowski, Abe Comfort, Hermann Muller, Joshua Lederberg, J. B. S. Haldane, Hudson Haogland, and Brock Chisholm.

16. From *Man and His Future*.

17. James Bonner, *The Next Billion Years: Mankind's Future in a Cosmic Perspective* (Moffett Field, Calif.: National Aeronautics and Space Administration, Ames Research Center, 1974).

18. A ludicrous example of the reductionist approach follows. Someone has seriously suggested that the efficiency of communications between individuals could be greatly increased if electrical connections were made between the approrpriate neurons of the individuals. The message would be communicated instantly and directly to the brain, without the cumbersome mouth-to-ear pathway. Whether or not such an electronic device is feasible is irrelevant. That an otherwise intelligent and thoughtful person could consider that an electronic device might improve on the human language—the very vehicle of thought—shows how far astray he has been led by a blind faith in technology. How often are new efforts in scientific research launched in this way? A little thought would have reminded him that language generates variety and new meaning through its very ambiguity; it enlarges our thought. Bohm has commented on this aspect of language:

In such a dialogue, when one person says something, the other person does not as a rule respond with exactly the same meaning as that seen by the first person. Rather, the meanings are only similar and not identical. Thus, when the second person replies, the first person sees a difference between what he meant to say and what the other person understood. On considering this difference, he may then be able to see something new, which is relevant both to his own views and to those of the other person. So it can go back and forth wit the continual emergence of a new content that is common to both participants. Thus, in a dialogue, each person does not attempt mainly to convey to the other certain ideas that are already known to him. Rather it may be said that the two people are participating in a single cyclical movement in which they are creating something new together.

David Bohm, W. Fuller, ed., in *The Biological Revolution*, (Garden City, N.Y.: Doubleday Anchor, 1970), pp. 36, 37.
19. *Harvard Magazine*, October 1976.
20. Ecologists H. T. and E. P. Odum have been a source of inspiration and guidance for the New Alchemists. The New Alchemists have given much thought to the world food supply, energy consumption, and population distribution. Using the scientific and philosophic foundation provided by H. T. Odum, they have constructed models for small-scale food production which are the antithesis of agribusiness. A few details of the workings of agribusiness show a stark contrast to those of ecological farming. Agribusiness is characterized by energy-intensive, inefficient, and costly procedures; small-scale farming is labor-intensive and puts human aspects back into food production. The public's view of agriculture is distorted for many reasons, but importantly by the physical character of the final product, neatly trimmed and packaged, sealing off the last vestiges of earth smells and sometimes colors. Moreover, fruits and vegetables that should exhibit natural variability look as if they had been produced by a computer. They have been grown from seeds treated chemically to preserve them and in soil given appropriate treatment with chemical fertilizers, fungicides, plants-insecticide, etc. The produce so obtained is packaged in refrigerated containers and transported to faraway places. The consumer does not know that agribusiness has taken over all these steps and executes them by means of energy-intensive methods; agribusiness would not be possible without fossil fuels. Incidentally, we know that the process is inefficient; it takes 5 to 20 calories of energy to produce the equivalent of 1 calorie of food. Consumers are therefore not aware that we are using vast amounts of nonrenewable resources in food production and that the inevitable consequence is pollution of the earth.

Worldwide, the export of food has had economic and social effects, particularly on underdeveloped nations; they have become dependent on items produced by the industrial community that they can ill afford, as well as having their agrarian potential disrupted. They are discouraged from growing native crops.

Agribusiness is not just a matter of planting and harvesting. Its various activities penetrate into every social, economic, and hence political domain. Virtually no area is untouched. A reversal of this big business to small farming holds the promise of eliminating the ill effects that inevitably accompany agribusiness.

21. Severo Ochoa, *New York Times,* February 3, 1980, p. B6.

Index

Abolition of Man, The (Lewis), 141
Adam, 5
Adenovirus-2, 84
"Affirmation of Freedom of Inquiry and Expression, An," 137
"Agent Orange," 95
American Civil Liberties Union, 120
Amniocentesis, 150
Anderson, E. S., 53
Ascot Conference, 54
Asilomar Conference, 58, 85, 102, 125, 136
Atomic Energy Commission, 75

Bach, 7
Bacon, Francis, 33, 134, 135
Bacteriophages, 25-26
Beckwith, Jonathan, 150
Berg, Paul, 110, 124
Biogen, 62
Biomedical science, emergence of, 24
Boffey, Philip, 125
Bohr, Niels, 105
Bonner, James, 153
Boyer, Herbert, 54, 62, 67, 85-86, 87
Brain Bank of America, The (Boffey), 125
Brave New World (Huxley), 153
Bridgeman, Percy, 30
Bronowski, Jacob, 22
Brooks, Harvey, 136
Bush, Vannevar, 34-35

Califano, Joseph, 147

Carroll, Lewis, 1
Carter, Jimmy, 66, 90
Chain, Ernst, 30
Chang, S., 115-17
Chargaff, Erwin, 104
Chemical and Engineering News, 72
"China syndrome," 53
Ciba Foundation, 152, 153
Club of Rome, 146
Coalition for Responsible Genetic Research, 155
Cohen, Stanley, 114, 115-17, 122
Commoner, Barry, 95
Complementarity, principle of, 43-45, 105
Consciousness, 4
Convergence, 1-11
Cosmos, 1
Cournand, André, 21, 134
Crick, Francis H. C., 25, 27, 80-81, 104, 105, 106, 107
Curtiss, Roy, III, 53, 114-15, 117, 119, 120, 121

Davis, Bernard, 50
Descartes, René, 33, 108, 135
Delbrück, Max, 105
Denton, Harold, 52
Doll, Sir Richard, 67
Double Helix, The (Watson), 27
DNA, *see* Recombinant DNA technology

Dresden II nuclear reactor accident, 52
Dubarle, D., 22

Earth, 1
E. I. DuPont de Nemours, 75
Einstein, Albert, 1
Electron microscopes, 79
Eli Lilly and Company, 62, 67
Ellul, Jacques, 19
Endless Horizons (Bush), 35
Environmental Protection Agency
   (EPA), 92-93
Eugenics, 150-54
Evolution, 2, 5, 6

Faldow, S., 53
Falmouth Conference, 53, 117-20
Federal Register, 111
Federation of American Societies for
   Experimental Biology, 101
Feeling, 8
Food and Drug Administration
   (FDA), 69, 93, 101
Ford, Henry, 65
Francis of Assisi, 3
Franklin, R., 104
Fredrickson, Donald S., 113, 118-119
Freedom of inquiry, 128-41; atomic fis-
   sion and, 129; claimed as a right,
   132-33; history of the issue, 128-29;
   reevaluation of needs and. 137

Galileo, 33, 38, 139
Gamow, George, 106
Genentech, 62
General Electric Co., 63
Gene sequences in DNA, 44
Genesis, Book of, 9
Gene-splicing, see Recombinant DNA
   technology
Genetic screening, 150, 151-52
Gilbert, Walter, 72, 132
Givaudan, ICMESA, 94, 95
Glass, Bentley, 39, 139
God, invariability of, 9
Goodman, Howard, 86-87, 88, 132
Gorbach, Sherwood, 117-20, 121
Gordon Research Conferences, 106-7,
   110, 120

"Grantsmanship," 28, 109
GRAS Substances, Select Committee
   on, 99-100
Green, Harold, 131, 136, 137, 138
Guillemin, Roger, 27

Haldane, J. B. S., 152
Halvorson, H. O., 120, 122
Handler, Philip, 110, 139
Heilbroner, Robert, 143
Hesiod, 9
Hoffman-La Roche, 94
Holley, Robert, 27
Holliday, Robin, 50, 53
Holman, Halstead, 54-55
Homeostasis, 41
Hopson, Janet L., 54
Hubbard, Ruth, 68
Hubris, cultural, 2
Human being, making of a, 6
Huxley, Julian, 152

Inquiry into the Human Prospect, An
   (Heilbroner), 143
Insulin, 63, 67-68, 73-73, 85, 89, 132
Intelligence, man's, 5
Interferon, 62

Karlesky, Joseph J., 37, 135
Kennedy, Edward M., 36, 113
Khorana, H. G., 27
Knowledge: original, 7; theory of, 6

Lear, John, 65, 89, 124, 144
Lederberg, Joshua, 57, 153
Leptophos, 92-93, 97
Lewis, Andrew M., Jr., 85
Lewis, C. S., 141
Limits to Growth, The, 146
Livermore, Shaw, 111
Luria, S. E., 127
Lysenko, Trofim, D., 38

Machines, operations of, 7
Manhattan Project, 75, 79
Matter, inorganic, 7
Merton, Robert, 89
Michelangelo Buonarroti, 5
Mislow, Kurt, 138

Möbius, Augustus, 13
Möbius Strip, 13-14
Molecular biology: basic understanding of, 42-43; potential power of, 107
Monod, Jacques, 22
Morphological descriptions, 81
Muller, Herman J., 153

National Academy of Sciences, 37, 71, 72, 99, 109, 110, 124-27, 135, 138
National Cancer Act of 1971, 29, 73
National Institute of Allergy and Infectious Diseases, 117
National Institute of Occupational Safety and Health, 92, 93
National Institutes of Health (NIH), 19, 36, 53, 60, 61, 67, 75, 77, 85-86, 87, 90, 91, 111, 112, 117, 123, 139; Guidelines for Recombinant DNA
Research, 18-19, 61, 103, 113, 123, 125
National Research Council, 125
National Science Foundation, 35, 75
Nature, 78, 94, 123
Nature: hierarchy of, 2; unitary structure of, 9
Nelson, Sen. Gaylord, 120
New Alchemists, 39, 156-57
New York Times, The, 66, 107
Newton, Isaac, 33, 89
Nirenberg, Marshall, 27
Nobel Prizes, 109, 110
Noll, Roger, 50
Nuclear Regulatory Commission (NRC), 52
Nylon, 75

Oak Ridge Research Reactor accident, 52
Out of the Night (Muller), 153

Pauling, Linus, 27, 105
PBB (polybrominated biphenyls), 95-98
Penicillin, 74
Philips Duphar, 94, 95
Physics, laws of, 7
Physiognomy, person's, 7

Plasmids, 47
Pollack, Robert, 110
President's Biomedical Research Panel, 35, 77
Principles, universal, 11
Proceedings of the National Academy of Sciences, 116

Rabi, I. I., 30
Recombinant DNA (Lear), 89
Recombinant DNA Advisory Committee (RAC), 123, 124
Recombinant DNA technology (gene-splicing): animal cells, 46-48; Asilomar Conference, 58, 85, 103, 125, 136; atmosphere of excitement generated by, 26-27; bacterial control systems, 41-42; base-pairing hypothesis, 105; biohazards, 59-60; Cambridge controversy, 112-13; cellular controls, 44-45; Central Dogma, described, 44-46; cloning, 152; commercial applications of, 49; competitive pressures in, 27-29; complimentarity, 43, 45, 105; control mechanisms and, 41-42; DNA defined, 44; discussed, 41-51; E. coli, 26, 53, 114-15, 117-19; early history of, 104-27; emergence of, 24-25; eugenics and, 150-54; food production and, 144-47; gene therapy, 144, 148-49; gene sequences, 44, 48; genetic code research, 27; genetic engineering risks, 49; Gordon Research Conferences, 106-7, 110, 120; "grantsmanship," 28; Harvard Biological Laboratories, controversy over, 112-13; Harvard Medical School and, 90-91; homeostasis, 41; hormone research, 27; hybrid DNA molecules, 47; insulin, 63, 67-68, 72-73, 85, 89, 132; interferon, 62; legislation for safety regulations, 60, 110, 112-24; lobbies, 60-61, 112-24; long-range consequences of, 15, 30, 103, molecules, most frequently studied, 43; National Academy of Sciences and, 124-27; National Institutes of Health, see National In-

Recombinant DNA technology (*continued*)

stitutes of Health (NIH); oil-eating bacteria, 63-65; organic irreversibility, 64-65; organism's need and DNA response, 42; plasmids, 47; protein product, 43; restriction enzymes, isolated, 26; risk-benefit analysis, 49-57; risks of large-scale application, 49; seduction of power and, 102-27; "shotgun" technique, 47-48, 50; societal implications of, 17, 32, 58, 91, 129; somatostatin, 62, 71-72; technological fixes and, 66-67; true needs, indentifying, 70; University of Michigan lab, 111-112; worldwide enthusiasm over, 131; *see also* Science; Technology
Rimbaud, Arthur, 152
Risk-benefit analysis, 39, 49-57, 137; cancer epidemic, 51, 53; economic view, 50; Falmouth Conference, 53; infections of the bloodstream, 54-55; mathematical view, 51; medical view, 50-51; nuclear power and, 51-53
Roszak, Theodore, 40, 82
Russell, Sen. Richard B., 34
Rutter, William, 85, 86, 87, 132

Saturday Review, 124
Schally, Andrew, 27
Schering-Plough and Inco, 73
Schmitt, Sen. Harrison H., 88, 89
Schwartz, Arthur, 51, 111
Science: advance in, 1; "antiscience" movements, 37; attitudinal changes, history of, 32-33; "blind emergence," 21; competition in research, 27-29, 72-73; computers and, 80; corporate state structure, 21, 134; Dioxin, 93-95; "disestablishment" of, 22; elitism in, 30; ethic, need for, 21; freedom, loss of, 76; freedom of inquiry, *see* Freedom of inquiry; freedom of technology and, 128; grants, peer-review system, 28, 109; human-

istic application of discoveries, 58; hybrid viruses, 85; idealization of scientists, 33; immediate goals, preoccupation with, 16; impersonality of research, 23; industrial chemicals, research in, 74-75; industry-oriented research, 38, 59, 62-64, 82; Leptophos, 92-93, 97; mission-oriented research, 16, 17, 31, 75-76, 77-78; myopic vision of, 131-32; National Cancer Act of 1971, 29, 73; national controls and, 22; Nobel Prizes, significance of, 109-10; politics and, 34-37, 103; post-World War II growth, 24-25, 33-35, 74-75; pragmatism, era of, 29; pressures on scientists to produce, 78; public interest groups, 155; recombinant DNA technology, *see* Recombinant DNA technology; reductionist vs. holistic view, 81, 129-30, 154; research instrumentation, 79-80; responsibility to public, 22-23, 31-33, 91, 97-98; secrecy in, 72; seduction of power and, 102-27; Select Committee on GRAS Substances, 99-100; "single vision" of, 40; social conscience, 16, 30-31, 37-38, 40, 83, 97-99, 142-57; social implications of research, 23, 30, 83, 155; social philosophy, lack of, 91; technicians, dependence upon, 80; as technology, 72-83; viral infection, study of, 81; *see also* Recombinant DNA technology; Technology
Science, 52, 71, 73, 90, 120, 121-122, 125
Scott, Robert, 95
Select Committee on GRAS Substances, 99-100
Senate Subcommittee on Science, Technology and Space, 51, 61, 71, 122
Shakespeare, William, 11
Siekevitz, Philip, 108
Silverstein, A. M., 36
Sinsheimer, R. L., 116
Smith, B. L. R., 37, 135
Smith, H. O., 26

Smith, H. W., 53
Smithsonian, 54
Somatostatin, 62, 71-72
Special Virus Cancer Program, 76-77
State of Academic Science, The (Smith and Karlesky), 135
Stevenson, Sen. Adlai, III, 85-86, 87-88, 89, 118, 122
Szybalski, Waclaw, 88

Targeted research, 75-76
Tempest, The (Shakespeare), 11
Technology: cancer and, 67; cyclic problems, 21; Dioxin case, 93-95; early recombinant DNA research and, 104-27; fixes for past failures, 66-67; as "the great healer," 22; hazards of success, 58-70; irreversibility of production, 65-66; economy based on, 59; Leptophos case, 92-93, 97; Michigan Chemical Corporation case, 95-98; problems caused by "solutions," 20; public disenchantment with, 17; risk-benefit analysis and, 39, 49-57; science as, 72-83; social responsibility in, 92-98; ultratechnology, 98; vast scale of, 39, 59; world food problem and, 69; *see also* Recombinant DNA technology; Science

"Test-tube" babies, 152
Thomas, Charles, 90, 91
Thought, human, 4
Three Mile Island accident, 52-53
Tydings, Sen. Millard E., 34

Union of Concerned Scientists, 155
Universe, 2; destiny of, 9
Upjohn Company, 62
Urey, Harold C., 31

Velsicol Corporation, 92-93, 97

Wade, Nicholas, 73, 90, 125
Wald, George, 112
Washington Post, The, 116, 120
Watson, James D., 25, 27, 80-81, 104, 105, 106
Whitehead, Alfred North, 9
Whirlwind, voice in, 1L
Wilkins, Maurice H. F., 104
Wilson, C. F., 37
World Health Organization (WHO), 69, 82, 92
World view, scientific, 2
Wright, Susan, 111

Zetlin, Fay, 14
Zinder report, 76

ABOUT THE AUTHOR

LIEBE F. CAVALIERI is a Member of the Sloan-Kettering Institute for Cancer Research, where his research has covered a broad spectrum of subjects relating to the molecular biology of DNA, especially DNA structure and enzymology. He is a Professor of Biochemistry in The Graduate School of Medical Sciences of Cornell University Medical College, where he headed the graduate program of the Sloan-Kettering Division for many years. Dr. Cavalieri received his doctorate in chemistry at the University of Pennsylvania and carried on additional studies at The Ohio State University, Columbia University and L' Institut de Recherches Scientifiques sur le Cancer, France.

In recent years Dr. Cavalieri has become concerned about the impact of science and technology on human society and the environment. He has spoken on this subject at a variety of public forums and has published articles in the *Southern California Law Review*, *The Poynter Center Essays on Recombinant DNA*, and the public press.

ABOUT THE FOUNDER OF THIS SERIES

RUTH NANDA ANSHEN, philosopher, author, editor, has founded, plans and edits WORLD PERSPECTIVES, RELIGIOUS PERSPECTIVES, CREDO PERSPECTIVES, PERSPECTIVES IN HUMANISM, THE SCIENCE OF CULTURE SERIES, THE TREE OF LIFE SERIES and CONVERGENCE. She also writes and lectures on the relationship of knowledge to man's nature and to the understanding of his place in society and in the cosmic scheme, since as man is in the universe so is the universe in man. Dr. Anshen's book, *The Reality of the Devil; Evil in Man,* a study in the phenomenology of evil, is published by Harper and Row. Dr. Anshen is a member of the American Philosophical Association, The International Philosophical Association, The History of Science Society and The Metaphysical Society of America.